NW Europe in Transition

The Early Neolithic in Britain and South Sweden

Edited by

Mats Larsson
Jolene Debert

BAR International Series 2475

2013

Published in 2016 by
BAR Publishing, Oxford

BAR International Series 2475

NW Europe in Transition

ISBN 978 1 4073 1087 9

BAR Publishing is the trading name of British Archaeological Reports (Oxford) Ltd.
British Archaeological Reports was first incorporated in 1974 to publish the BAR
Series, International and British. In 1992 Hadrian Books Ltd became part of the BAR
group. This volume was originally published by Archaeopress in conjunction with
British Archaeological Reports (Oxford) Ltd / Hadrian Books Ltd, the Series principal
publisher, in 2013. This present volume is published by BAR Publishing, 2016.

Printed in England

PUBLISHING

BAR titles are available from:

BAR Publishing
122 Banbury Rd, Oxford, OX2 7BP, UK
EMAIL   info@barpublishing.com
PHONE   +44 (0)1865 310431
FAX   +44 (0)1865 316916
www.barpublishing.com

TABLE OF CONTENTS

Context

This book is concerned with the developments that followed on from the introduction of farming into Britain and Southern Scandinavia (Denmark and Southern Sweden), and the idiosyncratic social and cultural patterns that emerged as the revolutionary potential of the Neolithic was gradually realised. Fundamental to our comparison is a concern with the ways in which communities inhabit their landscapes. If the Neolithic involved the introduction of new species of plants and animals and new forms of material culture into indigenous contexts, the longer-term consequences of this development should be gauged through changing practices of dwelling: patterns of occupation and mobility, the organisation of space, the location of ritual activities, the dead, and the sacred; and degrees of impact in ecological conditions.

The authors examine the implicit knowledge, habitual practice and material culture as forms of cultural inheritance which are passed between generations, and modified by innovation. In addressing this process it is important to recognise that artefacts, architecture and landscape are not merely the outcomes of human action, but the media through which human projects are carried forward. Of particular importance in this respect is the way in which people knowingly negotiate new identities and integrate new customs through reference to the past, in the form of social memory, material traces, and landmarks.

Mats Larsson

Chapter 1
In Dialogue:
a Condensed Outline of Debates on the Mesolithic – Neolithic Transition

Ellen McInnes

University of Manchester

Abstract: *This paper will briefly synthesise the history of how narratives about the Mesolithic – Neolithic transition in Britain have changed with the development of archaeology as a background to the papers within this volume. I will highlight how past shifts in archaeological thought have impacted these debates and altered the focus of different approaches. This particularly includes a focus on recent developments in Mesolithic research and how this might contribute to future discussions. By exploring this background I hope to expose the assumptions that have dominated, consider their continued influence today and provide a platform for future work, which begins with this volume. Introduction*

Keywords: *Culture-history, environment, radiocarbon dating, Childe, evolutionary theory, adaptation, transition*

Introduction

This paper aims to provide a brief overview of the development of ideas about the Mesolithic- Neolithic transition in Britain as a background to the papers in this volume. Rather than review current debates that will be discussed elsewhere, I will highlight how past shifts in archaeological thought have impacted these debates and altered the focus of different approaches and their continued influence today. Although much discussed (e.g. Childe 1949; Case 1969; Sherratt 1995; Whittle 1997; Sheridan 2000; Rowley-Conwy 2004; Thomas 2007; Cummings and Harris 2011) the Mesolithic – Neolithic transition in Britain remains poorly understood. The elusive nature of the transition in Britain has resulted in extensive debate about the change from Mesolithic lifeways to those of the Neolithic. Although recently have more precise dates become available (e.g. Whittle, Barclay *et al.* 2007; Whittle *et al.* 2011) a more detailed chronology only answers some of the questions.

It is widely acknowledged that archaeological investigation is carried out amid a variety of expectations and prejudices (Thomas 2004a, 1) and many of these continue to cast a shadow over discussions of the transition. By drawing out themes that persist in this debate, assumptions and prejudices that were inherent at the beginning, and during the development of ideas on this topic, can be laid bare. Challenges to these assumptions that have altered ideas about the transition, and the challenges that perhaps still need to be reinforced, will also be discussed.

It has been suggested that within the mass of literature on this topic a pattern has emerged of two opposing positions:

colonisation and indigenous adoption (Cummings and Harris 2011). This however, is a broad generalisation that ignores many of the more subtle models that discuss the transition. As problematic is the way in which the Mesolithic and Neolithic have been defined and placed in opposition. This is a theme that this paper will highlight within a historical discussion of how this dichotomy has been underlined by the different approaches taken to the transition and by those working in the two periods. The origins of this polarization of Mesolithic and Neolithic societies lie in the 19th century development of archaeology and studies of social change.

19th Century

During the late 19th Century ethnographers and evolutionary theorists (e.g. Spencer 1897; Spencer and Gillen 1899, Tylor 1871, 1889) suggested schemes and sequences of progression. These were based upon observations of contemporary societies that were assumed to represent different stages of a universal progression towards the civilised state of western Europe. Morgan (1877), for example, suggested three main stages: savagery; barbarism; and civilization, each with a number of sub-divisions. The stage that a society had reached was measured by the form of different aspects of society (Table 1.), such as technology, subsistence method and system of government. This assumed the uniformity of human civilizations and that each society progressed through the same stages of development each of which took the same form in all places at all times.

Within these schemes and theories of progression cultural change was interpreted as predominantly being driven by

Table. 1. The 'growth of intelligence and progress of mankind' as illustrated by the discoveries and inventions that mark beginning of successive phases. (after Morgan 1877)

Period	Conditions Marking the Beginning of the Phase
I. Lower Status of Savagery	From the Infancy of the Human Race to the commencement of the next Period.
II. Middle Status of Savagery	From the acquisition of a fish subsistence and a knowledge of the use of fire.
III. Upper Status of Savagery	From the Invention of the Bow and Arrow.
1V. Lower Status of Barbarism	From the Invention of the Art of Pottery.
V. Middle Status of Barbarism	From the Domestication of animals on the Eastern hemisphere, and in the Western from the cultivation of maize and plants by Irrigation, with the use of adobe-brick and stone.
VI. Upper Status of Barbarism	From the Invention of the process of Smelting Iron Ore, with the use of iron tools.
VII. Status of Civilization	From the Invention of a Phonetic Alphabet, with the use of writing, to the present time.

advances in technology. The natural course of development for all societies was understood to involve progression from hunter-gatherer, a barbaric and crude way of life, to farming. Agriculture represented security, prosperity, stability and provided the opportunities that eventually led to civilisation. The suggestion that more advanced cultures were more evolved and cognitively superior, appealed to colonial attitudes of the time;

It is both a natural and a proper desire to learn… how barbarians finally attained to civilization; and why other tribes and nations have been left behind in the race of progress. (Morgan 1877 Preface)

This positivist, evolutionary view of societies was used to argue that the hunter-gatherers of nineteenth century Australia or southern Africa were incapable of making the leap to the next stage in development. European contact would therefore eventually benefit these groups, lifting them out of their barbaric way of life (Dennell 1983, 154).

Alongside this colonial attitude there became an increasing concern with identifying the origins of specific groups. It was assumed that past societies could be placed within these schemes and archaeology used these analogies to establish a chronological framework with a focus on technological milestones. It was assumed that the stage each society had reached would be identifiable in the archaeological record by evidence of tool use and subsistence practices. Archaeology allowed an investigation of origins and for the temporal relationship of these evolutionary sequences in different regions to be established. Studies tracing the development of particular peoples became important in the search for national identity and unity in newly established European countries such as Germany and those under threat, such as Denmark (Trigger 1989, 149). At this time of increased nationalism Archaeology became the tool through which the origins of groups, races and nations could be investigated.

Childe: Trade and Typology

In 1925 Childe published the first edition of 'The Dawn of European Civilization' in which the Neolithic of Europe marked the beginnings of civilization with the arrival of agriculture, pottery and monumental constructions. However, rather than simply studying an inevitable evolutionary progress Childe was concerned with how the unique character of Neolithic European cultures enabled them to adapt and develop innovations from the near east (Trigger 1980, 669). Childe suggested that the inherent 'western' qualities of these first civilised people were the same that led to the inventions and commercial markets of the 20th Century.

The previous focus on evolutionary phases had also led to an emphasis on identifying major changes within societies ignoring spatial differences and small-scale change. Childe, in contrast, used variations in material culture, burial practices and monuments to define discrete cultures within geographical and temporal limits. Once identified these bounded cultures were used to create a relative chronology based on the assumption that agriculture first developed in the near east and spread by diffusion and migration across Europe.

However, before discussing the origin, development and spread of a European Neolithic, Childe (1925, 1-21) considered the 'transitional cultures' of Europe. Rather than recognising a distinct Mesolithic this period was termed the 'epi-palaeolithic', with a meaning different to that which it holds today, and groups were considered to be survivors from the Palaeolithic (see Rowley-Conwy 1996 for a discussion of this). Following the tone set by evolutionary studies of progression this era was set in opposition to the incoming Neolithic. Peake (1928, 19) described pre-Neolithic Europe as; 'an uncivilised state into which civilization was introduced'. The indigenous people of Europe were described 'as savages, in a primitive condition with a lowly, un-progressive culture' and were considered 'mere food gatherers; helpless and dependent on Mother Nature' (Peake and Fleure 1927, 10-13). In contrast 'Neolithic groups had superior tools, domesticated animals, cultivated plants and were in a position of dominance over much of nature' (Childe 1925, 1). As will be discussed later these generalised portrayals of Mesolithic and Neolithic societies and the dichotomy established at the beginning of the 20th century, have persisted, to varying degrees, in discussions of the transition.

Childe did recognise a distinction between the transitional cultures and those of the Palaeolithic and rejected such a sharp contrast, in essence however, Childe's model depicted indigenous European groups as uninventive and unable to progress. For Childe, the developments that characterised the Neolithic must have been introduced from elsewhere and it was suggested that Neolithic culture was brought to Britain via long-distance maritime trade networks, the establishment of small trading settlements and the diffusion of culture (Childe 1925, 286-92; 1934, 301). Alongside colonisation, Childe suggested that indigenous peoples were imitating and adopting Neolithic culture through contact with traders and settlers. This contact did not result in them being immediately wiped out and anomalous sites in the Early Neolithic were attributed to surviving indigenous groups. Within this brief overview of Childe's approach to Mesolithic groups and the arrival of Neolithic culture two important points should be highlighted. The first is the topic of maritime movement and the second the small reference to the continuation of some groups of hunter-gatherers, albeit only for a short length of time. Both these topics still feature prominently in discussions of the transition (e.g. Thomas 2004b; Cummings and Harris 2011; Garrow and Sturt 2011)

Much of the work by culture historians focussed on defining cultures and their physical boundaries (Trigger 1989, 205). These studies worked at increasingly smaller-scales providing detail at a local level but often obscuring the main features of each period (Dennell 1983, 1). This emphasis on the identification of bounded cultures, rather than universal developmental stages in societies, resulted in narratives of prehistory that traced the movements and interactions of groups. An ongoing concern with the origins of ethnic groups led to work that sought to trace cultures through their material culture and to place them in chronological order (e.g. Childe 1925; 1929; 1932; Whelan 1937; Piggott 1954). Great effort was taken by Childe (1925) to suggest that the different forms of monuments and artefacts in Britain were evidence of the different continental cultures that had an impact on the British Neolithic. Childe saw influences from Scandinavia, the Baltic, the Atlantic coast and North-west Europe (Fig. 1.). Chambered long barrows and cairns were suggested to be derived from western Mediterranean influences, more specifically corbelled roofs reflected links to Almeria, passages featuring lateral recesses implied contact with traders from Sardinia and long barrows indicated links to Holland. Pottery in Scotland was said to most closely resemble the Amorican style, while vessels in England were suggested to correspond to Baltic and Danish types (Childe 1925, 289-91).

With a focus on subsistence and technology it was through typology that Childe traced the origins, spread and development of civilisation in Europe and the subsequent development of the Neolithic. This continued search for

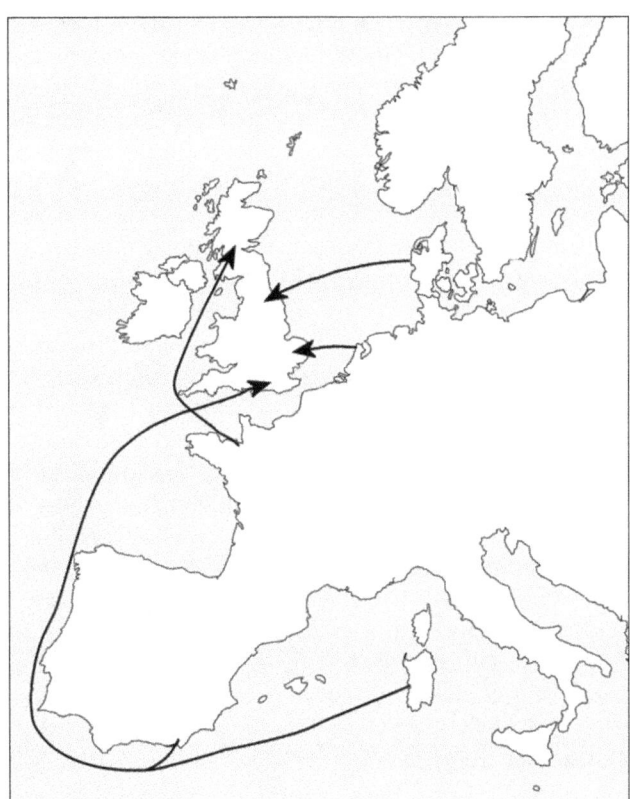

Fig. 1. Map showing examples of the continental influences Childe identified in the British Neolithic.

origins and an approach concerned with the movement of people and diffusion of ideas within a culture-historical framework rapidly became a common model for the transition to the Neolithic. For example, Peake (1928) took a similar economic approach to the development of the Neolithic. Like Childe, Peake suggested the arrival of civilisation in Britain was equated with the arrival of the Neolithic. Maritime and land trade routes complete with 'branch lines' (Peake 1928, 23), were suggested to be the path by which Neolithic culture reached Britain. The form and decoration of artefacts, especially pottery and stone tools, were used to determine contact and links between areas. Settlements of traders were thought to have introduced Neolithic technologies and subsistence practices to Europe and Britain. These Neolithic arts were then learnt by local populations, who imitated the material culture of the traders. In essence, from trading settlements Neolithic technologies spread to indigenous populations via a diffusion of ideas. As with the contrasting images created of Mesolithic and Neolithic societies these suggestions can still be found in current debates surrounding the transition in Britain. Sheridan, for example, (2000; 2003; 2004; 2007; 2010) puts forward a similar argument for colonisation and the acculturation of indigenous groups. Based on stylistic similarity and typological dating of pottery, monuments and timber structures, a movement of people from northern France to Britain is suggested to have occurred around 4000bc.

Invasion and colonisation driven by population expansion were favoured by culture-historians as explanations for the spread of the Neolithic, although the scale of migrations and the contribution of diffusion were debated. The Neolithic was understood to be primarily an economic phenomenon, involving subsistence and polished stone tools as part of a cultural package, whose development was borne of innate human desires for a more secure life. Work in the near-east however, led Childe (1954; 1958) to recognise that not all farmers were potters. This realisation that not all Neolithic traits were necessarily present in all places led more generally to an emphasis on the mode of subsistence as the key aspect of the Neolithic (Pluciennuk 1998, 63). A shift to agriculture and animal husbandry therefore came to be considered the fundamental change between a Mesolithic and Neolithic way of life, with other cultural elements of the Neolithic package being added onto this foundation. Migration, due to population increase, and subsequent diffusion of ideas remained the most common explanation for the expansion of the Neolithic despite the fact that in some areas evidence was yet to be found to support the theory (e.g. Barker 1975, 102).

Culture as Adaptation

Emerging approaches that were concerned with the relationship between societies and their environmental context tied into this new focus on mode of subsistence and resources, rather than the development of technology. This school of thought argued that innovations and cultural change came about in response to external changes in the environment. Childe (1950, 2) himself later recognised that a specific culture was an adaptation to a specific environment; a culture established in one environment would be likely to change significantly if transferred to another, a point that has been developed in post-processual approaches.

In 1939 Clark published the first edition of *Archaeology and Society*; in which he argued for a more functional approach to archaeology and outlined an ecologically influenced model of society based on an analogy of an organism within an ecosystem. A similar idea had briefly been mentioned by Childe (1936) but it was Clark that developed the idea of studying society as made up of systems developed as adaptive mechanisms in response to ecological constraints (Trigger 1989, 265). Culture was viewed as one of these adaptive systems and was suggested by Clark to facilitate co-operation between individuals.

Within his work Clark sought to investigate human development by studying changes in subsistence over time. Clark's (1954, 1972) excavations at Star Carr encouraged an approach that integrated the study of animal bone, pollen and other organic remains by palaeobotanists and zoologists in reconstructing subsistence patterns and social organisation. By investigating seasonal changes in local vegetation and climate and considering these alongside traces of activities and artefacts found at the site, Clark sought to document social life and subsistence patterns. This functional approach led to Clark's interpretations of Star Carr as a winter base from which a small group could hunt deer.

Within this approach culture change was suggested to come about as a response to variations within the relationship between society and its environment. Factors that could affect this were in essence similar to those that had been suggested by both culture-historians and those who had advocated evolutionary development (Trigger 1989, 269). However, the mechanism by which these causes brought about change differed. Technological innovations, population fluctuation, climate change and cultural contact were suggested to be external factors that might bring upset the equilibrium of the ecosystem within which environment and culture co-existed. Changes in lithic technology or pottery style could therefore be explained by changes in the environment, climate or demographic of an area. In an article criticising the over-dependence of archaeologists on an 'overseas influence' to explain change or developments in British prehistory Clark (1966) addressed the question of the beginning of the Neolithic. Whilst pottery styles or lithic technology could, he argued, be transferred between groups, subsistence, as a fundamental aspect of culture, could only be introduced through population movement (Thomas 1999, 445). The beginning of the Neolithic in Britain therefore could be attributed to colonisation with the change in subsistence as the primary and most important change to occur.

Clark's work formed part of a move away from typology in archaeology within a broader shift in archaeological approaches from an interest in technology to one that prioritised subsistence and economy. This new approach sought to link changes in the cultural environment, such as climate change, population pressure and resource failure, with economic, and eventually cultural, change. Higgs and Jarman (1975), following Clark, argued that all of human behaviour could be attributed to the mechanisms involved in balancing the relationship between population levels and the environment. Their development of methods such as site-catchment analysis allowed the environment and economic strategies to be investigated. For example, Williams (1989, 518) argued that Mesolithic groups in Britain adopted cereal cultivation to increase the level of carbohydrate in their diet. While these studies have been criticised as tending to 'study man as a bipedal stomach that ingests anything worth eating within an hour of his settlement' (Dennell 1983, vii), the impact of methods such as palaeoeobotany, zooarchaeology and bioarchaeology and the focus on environment and subsistence have continued to impact studies of the transition (e.g. Bonsall *et al.* 2002a; 2002b).

Culture as a means of adaption also continued to feature within archaeological literature. Binford (1965, 205) suggested that culture interacted with a number of other systems such as climate and demographics but stressed

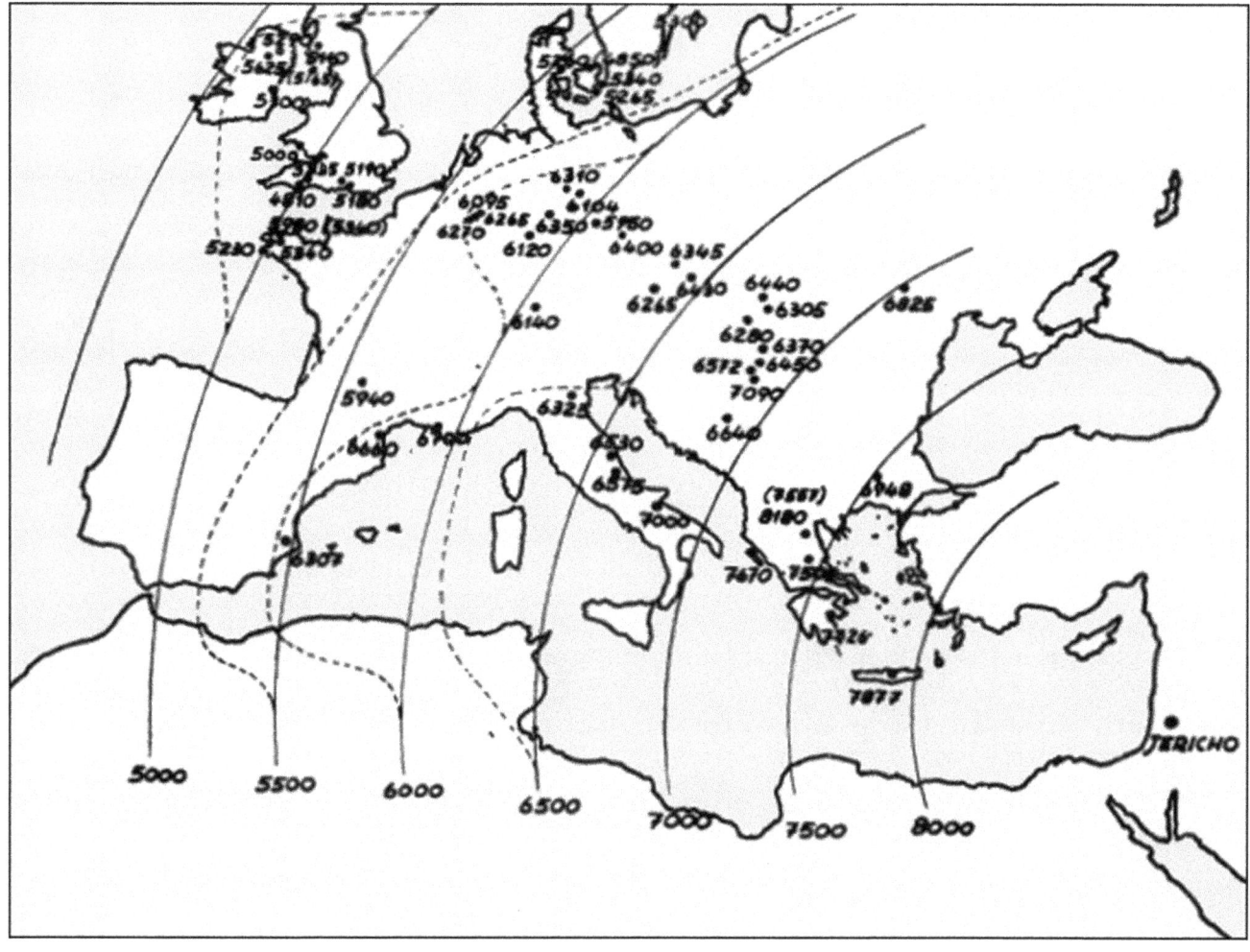

Fig. 2. The 'Wave of Advance' model measuring the spread of farming across Europe (Ammerman and Cavalli-Sforza 1971, 685)

that this was not a deterministic relationship. Instead he argued that culture was at the point where society and the environment meet, with cultural innovations a way of adapting to changes (Thomas 1999, 450). Within the new archaeology, climate change, population pressure and resource failure were repeatedly cited in the archaeological literature as causes of cultural change (e.g. Binford 1968; Meyers 1971; Flannery 1973; Cohen 1975; 1977). In contrast to culture-historical studies these new approaches led to the trend for detailed local studies being reversed with echoes of the late 19th century universal schemes in large-scale, thematic, generalised cross-cultural studies being used to investigate processes of change and adaptations to the environment.

Radiocarbon models

Ammerman and Cavalli-Sforza retained the focus on subsistence and discussed the diffusion of the Neolithic through Europe as synonymous with the spread of agriculture. Using radiocarbon dates their initial model measured the rate at which farming spread across Europe (Fig. 2.) and was followed by a similar study of population

growth (Ammerman and Cavalli-Sforza 1973). One cause was thought to be applicable for the European continent: sedentary farming was suggested to cause a rise in population, at a certain level local populations became unsustainable and caused the movement of people in an interrupted wave across Europe.

The validity of the model has been challenged with the reliability of the radiocarbon dates particularly criticised; some related to sites where agriculture was fully established and the main form of subsistence, whilst others were isolated finds of domesticated resources. In some instances dates were drawn from contexts in which evidence of agriculture was absent but other Neolithic material culture was present (Thomas 1996). It is only by considering the Neolithic to be a package of traits that these dates could be considered valid for their study. Work by Childe some decades earlier, mentioned above, makes this position difficult to justify. Further criticisms to this model came from the tentative beginnings of new approaches to archaeology that encouraged an interpretation beyond scientific measure. Models such as the 'wave of advance', that rely simply on the movement of people under population pressure as an explanation for change, have been heavily criticised as failing to address the social interactions that must have been involved in the Mesolithic – Neolithic transition. Instead they simply record the dates of the first appearance of a Neolithic trait. More generally, large-scale models have been suggested to be inappropriate for a transition that became recognised as having taken place in a diverse range of local contexts.

The need to understand these local transitions and social processes began to be addressed with new approaches that explored other mechanisms by which change might be brought about. In the western Mediterranean, for example, rather than colonisation it was suggested that earlier networks and connections between hunter-fisher-gatherer communities and Cardial Ware Neolithic groups later provided a pathway for the introduction of elements of the Neolithic (Lewthwaite 1981). In Scandinavia domesticates were adopted by indigenous populations over many centuries. Rather than the replacement of Late Ertebolle groups with incoming Neolithic farmers, Jennbert (1985) suggested that agriculture may have been introduced via networks of contact and exchange with continental Europe, involving gifts of cattle and the movement of people in marital alliances. Within this work the relationship between social change and changes in material culture is emphasised as crucial to understanding the conditions within which change occurred (Jennbert 1985, 196).

The idea of indigenous adoption was echoed in other accounts at this time. Price (1983) suggested that successful Mesolithic adaptations in north-west Europe caused a delay in the uptake of agriculture. Hunter-gatherer groups adopted and used pottery without needing to take up farming. Rather than invading farmers Price suggested that only a change in climate brought about the need for economic change within the context of existing societies. Studies such as these challenged the idea that the transition to the Neolithic simply involved the replacement of one group or culture with the next in an, all-encompassing, unidirectional manner. A focus on subsistence remained however, with models of the transition centred on the factors that might cause the adoption of agriculture and the mechanisms by which it occurred.

The explanation of climate change as a trigger for the adoption of agriculture is a theme that echoes earlier concerns with the environment and has repeatedly featured in discussions of the transition. Although this paper does not intend to cover current debates on the topic Rowley-Conwy (2004) and Bonsall *et al.* (2002a; 2002b) provide a good overview of this argument and recent research.

Zvelebil and Rowley-Conwy

As part of this shift from tracing migrations to considerations of social processes, Zvelebil and Rowley-Conwy (1984; 1986) proposed a three-stage model that was explicitly focussed on the adoption of agriculture.

The model suggested that the adoption of agriculture by hunter-gatherer communities progressed through stages of `availability', `substitution' and `consolidation' with the potential for each stage to last a different amount of time in different contexts. In each stage a different relationship to domesticated resources was established, with the implication that only when a community was dependent on agriculture were they Neolithic (Thomas 2007, 424). Unlike many previous studies Zvelebil and Rowley-Conwy (1984; 1986) explicitly addressed the motivation for indigenous populations to adopt agriculture with the dynamics of cultural contact considered as a cause of change (Pluciennik 1999, 666). They argued that a crisis in wild resources, or a perceived advantage to becoming farmers, must have driven the process echoing earlier studies of the relationship between man, the environment and change.

Although this model continued to implicitly define the Neolithic as a unified package, with changes in other aspects of culture secondary to, and tied to, an initial shift to agriculture, there was some consideration of the temporality and processes by which Mesolithic communities became Neolithic. The potential for each stage to last a different amount of time in different contexts, and the changing relationships between man and domesticated resources formed part of a more subtle discussion of the transition however; this model still considers the adoption of agriculture the primary change despite challenges to this assumption. The continued focus on subsistence meant that this model only discussed the transition in economic terms with other changes in material culture given less attention (Whittle 1997, 6). Whilst each stage could last for different lengths of time the model was suggested to be applicable across Europe leaving little room for the potentially varied responses of the Mesolithic communities of Europe.

A more thorough reanalysis of the Mesolithic-Neolithic transition came with the wider criticism of the idea of cultures as a package of traits, customs, technologies and economic strategies regularly occurring together. The idea that different aspects of the Neolithic may have been adopted separately and selectively by Mesolithic groups became more widely discussed and led to criticisms of the primacy given to agriculture. Despite this focussed discussion of the social construction of Neolithic societies in specific contexts with an emphasis on local conditions, the idea that Mesolithic communities were homogenous is often still implicit in writings on the transition.

Mesolithic Misunderstandings

Whilst the recent argument that we should not focus on debates of colonisation versus indigenous adoption (Cummings and Harris 2011) has identified one recurrent theme in the literature it is the dichotomy of Mesolithic and Neolithic societies that I wish to highlight here. Much of this divide stems from the early classifications of societies and cultures. Evolutionary theorists created an inferior stereotype of Mesolithic groups and this view of pre-Neolithic societies as primitive peoples just surviving with simple technologies has been perpetuated until recent decades. This view became further established through the continued focus on technology and subsistence as measures of sophistication and complexity. Dušan Borić (2002) has suggested that the forager and the farmer have been discussed as if they were different entities that interacted at the fringes of their two respective worlds. As well as being divided into two discrete time blocks this simplified dichotomy of subsistence practices, technology and lifeways has created social divides. This has been reflected in models of the transition that describe a shift between these discrete two categories with no room for the many variants that lie in-between, or within, foraging and farming ways of life.

Although discussions of the transition have developed to incorporate theories of agency, practice, social interaction, and a range of mechanisms of change, the way in which the two periods are placed in opposition remains one of the most fundamental problems in debates about the transition. The period is rarely covered evenly and the late Mesolithic and early Neolithic are too frequently studied in isolation with little communication.

In 1983 Dennell suggested that if indigenous adoption were to be considered there was a need for 'a much closer dialogue between Mesolithic and Neolithic prehistorians' and that the transition is not a 'convenient point at which studies of prehistoric hunter-gatherers can end, and those of agriculturalists can begin' (Dennell 1983, 189). It could be argued that this remains a valid plea. The suggestion that the Mesolithic – Neolithic transition be thought of as more a transformation of a range of worldviews and ways of life requires an approach that focuses neither on the Mesolithic nor Neolithic, and requires communication between archaeologists working in each period- something this volume aims to facilitate in order to move away from the assumptions that have dominated past debates.

References

Ammerman, A.J., Cavalli-Sforza, L. L. 1971. Measuring the rate of spread of early farming in Europe. *Man* 6, 674 - 688.
    1973. A population model for the diffusion of early farming in Europe. In C. Renfrew (ed.), *The Explanation of Culture Change: Models in Prehistory*, 343-358. London, Duckworth.
Barker, G. W. 1975. Early Neolithic Land Use in Yugoslavia. *Proceedings of the Prehistoric Society*. 41, 85-104.
Binford, L. R. 1965. Archaeological Systematics and the Study of Culture Process. *American Antiquity* 3, 1203-1210.
    1968. Post-Pleistocene adaptations. In S. R. Binford and L. R. Binford (eds.), *New perspectives in archaeology*, 21-49. Chicago, Aldine.
Bonsall, C., Macklin, M. G., Anderson, D. E. and Payton, R. W. 2002a. Climate change and the adoption of agriculture in northwest Europe. *European Journal of Archaeology* 5, 7–21.
Bonsall, C., Macklin, M.G., Payton, R.W. and Boronean, A. 2002b. Climate, floods, and river gods: Environmental change and the Meso-Neolithic transition in southeast Europe. *Before Farming* 3, 1-12.
Borić, D. 2002. The Lepenski Vir conundrum: reinterpretation of the Mesolithic and Neolithic sequences in the Danube Gorges. *Antiquity* 76, 1026-1039.
Case, H. 1969. Neolithic Explanations. *Antiquity* 43, 176-86.
Childe, V. G. 1925. *The Dawn of European Civilization*. London, Kegan, Paul.
    1929. *The Danube in Prehistory*. Oxford, Oxford University Press.
    1932. Chronology of Prehistoric Europe. A Review. *Antiquity*, 206-212.
    1934. *New Light on the Most Ancient East: The Oriental Prelude to European Prehistory*. London, Kegan, Paul.
    1936. *Man Makes Himself*. London, Collins.
    1949. The Origin of Neolithic Culture in Northern Europe. *Antiquity* 23, 129-135.
    1950. *Prehistoric Migrations in Europe*. Oslo, Aschehaug.
    1954. *New Light on the Most Ancient East. Revised Edition*. London, Kegan, Paul.
    1958. *The Prehistory of European Society*. London, Penguin Ltd.
Clark, J. G. 1954. *Excavations at Star Carr. An Early Mesolithic Site at Seamer, near Scarborough, Yorkshire*. Cambridge, Cambridge University Press.
    1966. The invasion hypothesis in British prehistory. *Antiquity* 40, 172 -189.
    1972. *Star Carr: A Case Study in Bioarchaeology. Addison-Wesley Modules in Archaeology*. New York, Addison-Wesley.
Cohen, M. R. 1975. Population pressure and the origin of agriculture. In S. Polgar (ed.), *Population, ecology and social evolution*, 79 – 121. The Hague, Mouton.
    1977. *The food crisis in prehistory: Over population and the origins of agriculture*. New Haven, Yale University Press.
Coon, C. S. 1939. *Races of Europe*. New York, Macmillan.
Cummings, V. and Harris, O. 2011. Animals, People and Places: The Continuity of Hunting and Gathering Practices across the Mesolithic-Neolithic Transition in Britain. *European Journal of Archaeology* 14(3), 361 - 382.
Dennell, R. 1983 *European Economic Prehistory: a new approach*. London, Academic Press.
Flannery, K. 1973. The origins of agriculture. *Annual Review of Anthropology* 2, 271-310.
Garrow, D. and Sturt, F. 2011. Grey waters bright with Neolithic Argonauts? Maritime connections and the Mesolithic-Neolithic transition within the 'western seaways' of Britain, c. 5000–3500 BC. *Antiquity* 85, 59–72.

Higgs, E.S. and Jarman, M. R. 1975. Palaeoeconomy. In E. S. Higgs. (ed.), *Palaeoeconomy,* 1-8. Cambridge, Cambridge University Press.

Jennbert, K. 1985 Neolithization: a Scanian persepective. *Journal of Danish Archaeology* 4, 196-197.

Lewthwaite, J. 1981. Ambiguous first impressions: a survey of recent work on the early Neolithic of the west Mediterranean. *Journal of Mediterranean Anthropology and Archaeology* 1, 292-307.

Meyers, J. T. 1971. The origins of agriculture: An evaluation of three hypotheses. In S. Streuver (ed.) Prehistoric agriculture, 101-21. Garden City, Natural History Press.

Morgan L. H.1877. *Ancient Society.* London, MacMillan and Company.

Peake, H. J. E. 1928 Presidential Address. The Introduction of Civilization into Britain. *Journal of the Royal Anthropological Institute of Great Britain and Ireland* 58, 19-31.

Peake, H. J. E. and Fleure, H. J. 1927. *Peasants and Potters.* Oxford, Oxford University Press.

Piggott, S. 1954. *The Neolithic cultures of the British Isles.* Cambridge, Cambridge University Press.

Pluciennuk, M. 1998. Deconstructing 'The Neolithic'; in the Mesolithic-Neolithic transition. In M. Edmonds and C. Richards (eds.) *Understanding the Neolithic of north-western Europe,* 61-83. Glasgow, Cruithne Press.

Price, T.D. 1983. The European Mesolithic. *American Antiquity* 48(4): 761-778.

Rowley-Conwy, P. 1996. Why didn't Westrop's 'Mesolithic' catch on in 1872?. *Antiquity* 70, 940-944.

2004. How the West Was Lost. A Reconsideration of Agricultural Origins in Britain, Ireland and Southern Scandinavia. *Current Anthropology* 45, S83 – S109.

Sheridan, A 2000 Achnacreebeag and its French connections: Vive the 'Auld Alliance'. In J. Henderson (ed.) *The Prehistory and early history of Atlantic Europe,* 1-15. Oxford, British Archaeological Reports International Series 861.

2003. Ireland's earliest 'passage' tombs: a French connection? In G. Burenhult (ed.), *Stones and bones: formal disposal of the dead in Atlantic Europe during the Mesolithic-Neolithic interface 6000-3000 BC,* 9-27. Oxford, British Archaeological Reports International Series 1201.

2004. Neolithic connections along and across the Irish Sea. In V. Cummings and C. Fowler (eds.), *The Neolithic of the Irish Sea: Materiality and Traditions of Practice,* 9–21. Oxford, Oxbow.

2007. From Picardie to Pickering to Pencraig Hill? New information on the 'Carinated Bowl Neolithic' in northern Britain. In A. Whittle and V. Cummings (eds.), *Going Over: The Mesolithic Neolithic Transition in North West Europe,* 41–492. London, British Academy.

2010. The Neolithization of Britain and Ireland: the 'big picture'. In B. Finlayson and G. Warren (eds.), *Landscapes in Transition,* 89–105. Oxford, Oxbow.

Sherratt, A. 1995. Instruments of Conversion? The Role of Megaliths in the Mesolithic/Neolithic Transition in North-West Europe. *Oxford Journal of Archaeology* 14, 245-260.

Spencer, H. 1897. *Principles of Sociology.* 3 Volumes. New York, Appleton & Co.

Spencer, W. B. and Gillen, F. J. 1899. *The Native tribes of central Australia.* London, MacMillan & Co.

Thomas, J.S. 1996. The cultural context of the first use of domesticates in continental Central and Northwest Europe. In Harris, D. (ed.), *The origins and spread of agriculture and pastoralism in Eurasia,* 310-322 London, UCL Press.

1999. Culture and Identity. In Barker, G. (ed.) *Companion encyclopaedia of archaeology,* 431-469. London, Routledge.

2004a *Archaeology and Modernity.* London, Routledge.

2004b. Current debates on the Mesolithic-Neolithic transition in Britain and Ireland. *Documenta Praehistorica* 31, 113–130.

2007. Mesolithic-Neolithic transitions in Britain: from essence to inhabitation. In A. Whittle and V. Cummings (eds.), *Going Over: The Mesolithic-Neolithic Transition in Europe,* 423-440. London, British Academy.

Trigger, B. G. 1980. Archaeology and the Image of the American Indian. *American Antiquity,* 45 (4), 662 -676.

1989. *A History of Archaeological Thought.* Cambridge, Cambridge University Press.

Tylor, E. B. 1871. *Primitive Culture.* London, Murray.

1889 On a Method of Investigating the Development of Institutions Applied to Laws of Marriage and Descent. *Journal of the Royal Anthropological Institute* 18, 245-269.

Whelan, C. B. 1937. Studies in the Significance of the Irish Stone Age: The Culture Sequence. *Proceedings of the Royal Irish Academy* 44, 115-137.

Williams, E. 1989. Dating the introduction of food production into Britain and Ireland. *Antiquity* 63, 510–521.

Whittle, A.W.R. 1997 *Europe in the Neolithic: the creation of new worlds.* Cambridge, Cambridge University Press.

2007. The temporality of transformation: dating the early development of the southern British Neolithic. In Whittle, A. and V. Cummings (eds.), *Going Over: The Mesolithic-Neolithic Transition in Europe,* 377-398. London, British Academy.

Whittle, A., Barclay, A., Bayliss, A., McFadyen, L., Schulting, R. and Wysocki, M. 2007. Building for the Dead: Events, Processes and Changing Worldviews from the Thirty-eighth to the Thirty-fourth Centuries cal. bc in Southern Britain. *Cambridge Archaeological Journal* 17(1), 123–147.

Whitte, A., Healy, F. and Bayliss, A., (eds.). 2011. *Gathering Time: dating the early Neolithic enclosures of southern Britain and Ireland.* Oxford, Oxbow Books.

Zvelebil, M. and Rowley-Conwy, P. 1984. Transition to farming in northern Europe: a gatherer-hunter perspective. *Norwegian Archaeological Review* 17, 104–127.

1986. Foragers and farmers in Atlantic Europe. In M. Zvelebil. (ed.), *Hunters in transition: new directions in archaeology,* 67-89. Cambridge, Cambridge University Press.

CHAPTER 2

FARMING NEW LANDS IN THE NORTH:
THE EXPANSION OF AGRARIAN SOCIETIES DURING THE EARLY NEOLITHIC IN SOUTHERN SCANDINAVIA

Lasse Sørensen

Abstract: *In this paper I will argue, through a series of 14C dates with primary evidence of agriculture, that the expansion of agrarian societies towards Southern Scandinavia was a swift process occurring from 4000 to 3700 BC. The expansion involved the migration of smaller groups of pioneering farmers originating from Middle Neolithic communities in Central Europe, thus ruling out local Ertebølle hunter-gatherers as the primary carriers of agrarian technology and ideology. The pioneering farmers brought with them a complete agrarian technology, new material culture and structures. The reason for the expansion is still uncertain, but a growing population pressure combined with the fact that cultivation by slash and burn method requires so much space, that it could force some pioneering farmers to cultivate new lands in The North. The transition towards an agrarian way of life probably occurred through a complex and continuous process of migration, integration and gradual assimilation between pioneering farmers and local hunter-gatherers.*

Keywords: *Southern Scandinavia, Neolithisation, Immigration, Ideology, Agriculture, Expansion and Transition.*

Introduction

The objective of this paper is to investigate how fast the expansion of agrarian societies occurred in Southern Scandinavia (Northern Germany, Denmark and southern and western Sweden) during the Early Neolithic (4000-3500 BC). We have to acknowledge the fact, that there are problems when using 14C dates from the transition between the Late Mesolithic and Early Neolithic. The problem stems from two wiggles which have been observed on the calibration curve. The first one is located from 4200 to 4050 cal BC and the second one from 3950 to 3790 cal BC; thus limiting resolution and creating some division in our C-14 data (Reimer *et al.,* 2009). Nevertheless, the entrenched discussion of whether agriculture was introduced by migrating agrarian societies or indigenous populations could become a combination of the two hypotheses, because the transition towards agriculture happened at different speed in each region. Indigenous hunter-gatherers became farmers within one or two generations in some regions, thus playing an important role in the spreading of the agrarian way of life to other hunter-gatherer tribes. In other regions the transition was more of a slow process, lasting several generations. Within this transitional process it becomes important to discuss what separates a farmer from a hunter-gatherer.

Definition of farmers and hunter-gatherers

In my opinion hunting, gathering and fishing is practiced by both hunter-gatherer and farmers. What separates farmers from hunter-gatherers in a transitional context is crop cultivation and managing husbandry all year round. Firstly, cultivation requires a whole new set of technology including slash and burn activities for opening the landscape, preparing fields, sowing and growing crops, grain processing and storing seeds. Secondly, keeping domesticated animals all year round requires storage of food for the winter. However, I do not see any problem with the fact, that Late Mesolithic or Early Neolithic hunter-gatherers could have kept a few domesticated animals for meat reserves and prestige reasons. The managing of a few domesticated animals could be interpreted as initial herding activities by communities that still live mainly as hunter-gatherers.

Cereal grains

Direct dating of charred cereal grains of Emmer wheat (*Triticum dicoccoides*), Einkorn wheat (*Triticum monococcum*) and Naked barley (*Hordeum vulgare convar nudum*) at Early Neolithic sites in Southern Scandinavia places them between 4000 and 3800 BC. Pollen analyses also shows that this period is synchronic with a higher degree of Ribwort plantain (*Plantago lanceolata*) and Birch (*Betula*). This could indicate slash and burn forest clearance (Andersen 1993, 161; Odgaard 1994, 1; Rasmussen 2005, 1116; Sjögren 2006) (Fig. 1). At the same time, a few grain impressions have been interpreted in some Late Ertebølle potsherds from the coastal sites of Löddesborg and Vik in Scania (Jennbert 1984). However, both sites contain intermixed layers with Late Ertebølle and Early Funnel beaker ceramics. Therefore, it is possible that these sherds originated from funnel beakers, as they have the same coarse tempering and thickness as the Ertebølle ceramics (Koch 1987, 107). Currently, there is no other archaeological evidence supporting any kind of cultivation during the Late Mesolithic in Southern Scandinavia.

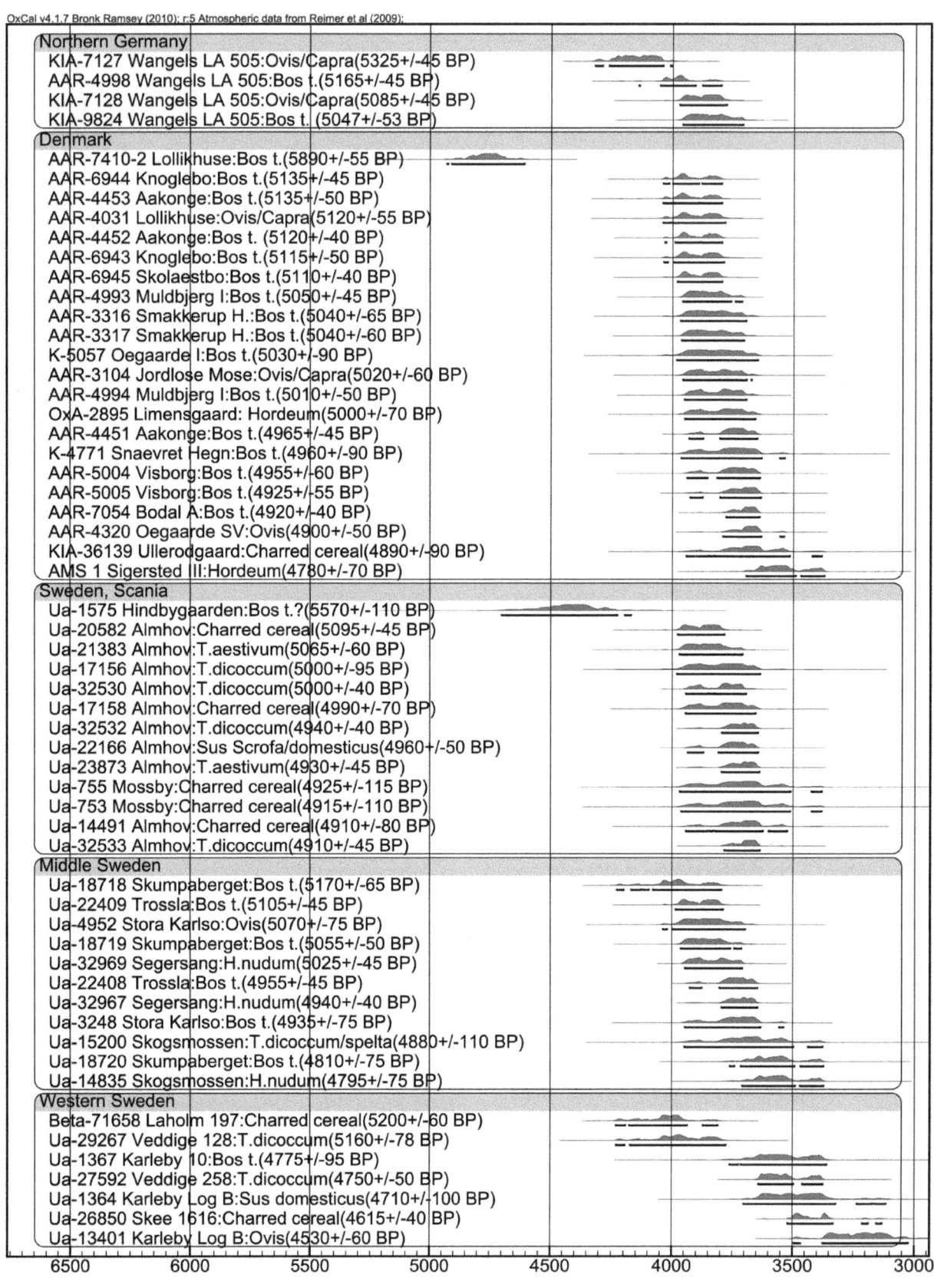

OxCal v4.1.7 Bronk Ramsey (2010); r:5 Atmospheric data from Reimer et al (2009);

Northern Germany
KIA-7127 Wangels LA 505:Ovis/Capra(5325+/-45 BP)
AAR-4998 Wangels LA 505:Bos t.(5165+/-45 BP)
KIA-7128 Wangels LA 505:Ovis/Capra(5085+/-45 BP)
KIA-9824 Wangels LA 505:Bos t. (5047+/-53 BP)

Denmark
AAR-7410-2 Lollikhuse:Bos t.(5890+/-55 BP)
AAR-6944 Knoglebo:Bos t.(5135+/-45 BP)
AAR-4453 Aakonge:Bos t.(5135+/-50 BP)
AAR-4031 Lollikhuse:Ovis/Capra(5120+/-55 BP)
AAR-4452 Aakonge:Bos t. (5120+/-40 BP)
AAR-6943 Knoglebo:Bos t.(5115+/-50 BP)
AAR-6945 Skolaestbo:Bos t.(5110+/-40 BP)
AAR-4993 Muldbjerg I:Bos t.(5050+/-45 BP)
AAR-3316 Smakkerup H.:Bos t.(5040+/-65 BP)
AAR-3317 Smakkerup H.:Bos t.(5040+/-60 BP)
K-5057 Oegaarde I:Bos t.(5030+/-90 BP)
AAR-3104 Jordlose Mose:Ovis/Capra(5020+/-60 BP)
AAR-4994 Muldbjerg I:Bos t.(5010+/-50 BP)
OxA-2895 Limensgaard: Hordeum(5000+/-70 BP)
AAR-4451 Aakonge:Bos t.(4965+/-45 BP)
K-4771 Snaevret Hegn:Bos t.(4960+/-90 BP)
AAR-5004 Visborg:Bos t.(4955+/-60 BP)
AAR-5005 Visborg:Bos t.(4925+/-55 BP)
AAR-7054 Bodal A:Bos t.(4920+/-40 BP)
AAR-4320 Oegaarde SV:Ovis(4900+/-50 BP)
KIA-36139 Ullerodgaard:Charred cereal(4890+/-90 BP)
AMS 1 Sigersted III:Hordeum(4780+/-70 BP)

Sweden, Scania
Ua-1575 Hindbygaarden:Bos t.?(5570+/-110 BP)
Ua-20582 Almhov:Charred cereal(5095+/-45 BP)
Ua-21383 Almhov:T.aestivum(5065+/-60 BP)
Ua-17156 Almhov:T.dicoccum(5000+/-95 BP)
Ua-32530 Almhov:T.dicoccum(5000+/-40 BP)
Ua-17158 Almhov:Charred cereal(4990+/-70 BP)
Ua-32532 Almhov:T.dicoccum(4940+/-40 BP)
Ua-22166 Almhov:Sus Scrofa/domesticus(4960+/-50 BP)
Ua-23873 Almhov:T.aestivum(4930+/-45 BP)
Ua-755 Mossby:Charred cereal(4925+/-115 BP)
Ua-753 Mossby:Charred cereal(4915+/-110 BP)
Ua-14491 Almhov:Charred cereal(4910+/-80 BP)
Ua-32533 Almhov:T.dicoccum(4910+/-45 BP)

Middle Sweden
Ua-18718 Skumpaberget:Bos t.(5170+/-65 BP)
Ua-22409 Trossla:Bos t.(5105+/-45 BP)
Ua-4952 Stora Karlso:Ovis(5070+/-75 BP)
Ua-18719 Skumpaberget:Bos t.(5055+/-50 BP)
Ua-32969 Segersang:H.nudum(5025+/-45 BP)
Ua-22408 Trossla:Bos t.(4955+/-45 BP)
Ua-32967 Segersang:H.nudum(4940+/-40 BP)
Ua-3248 Stora Karlso:Bos t.(4935+/-75 BP)
Ua-15200 Skogsmossen:T.dicoccum/spelta(4880+/-110 BP)
Ua-18720 Skumpaberget:Bos t.(4810+/-75 BP)
Ua-14835 Skogsmossen:H.nudum(4795+/-75 BP)

Western Sweden
Beta-71658 Laholm 197:Charred cereal(5200+/-60 BP)
Ua-29267 Veddige 128:T.dicoccum(5160+/-78 BP)
Ua-1367 Karleby 10:Bos t.(4775+/-95 BP)
Ua-27592 Veddige 258:T.dicoccum(4750+/-50 BP)
Ua-1364 Karleby Log B:Sus domesticus(4710+/-100 BP)
Ua-26850 Skee 1616:Charred cereal(4615+/-40 BP)
Ua-13401 Karleby Log B:Ovis(4530+/-60 BP)

6500   6000   5500   5000   4500   4000   3500   3000

Calibrated date (calBC)

Fig. 1. C-14 dates showing the expansion of agrarian societies during the Early Neolithic. All radiocarbon dates have been calibrated using the OxCal v4.1.7 program. Data after: (Hartz and Lübke 2005, 119ff; Sørensen 2005, 304f; Heinemeier 2002, 273f; Heinemeier and Rud 1998, 303; 1999, 340; 2000, 302; Price and Gebauer 2005, 123; Koch 1998, 253; Fischer 2002; Hadevik 2009; Rudebeck 2010, 112ff; Larsson 1992, 74; Hallgren 2008; Lindqvist and Possnert 1997; Svensson 2010; Johansson et al. 2011; Persson 1999; Ryberg 2006; Westergaard 2008; Esben Aasleff personal communication; Karl Göran Sjögren personal communication).

Domesticated animals

Domesticated cattle (*Bos Taurus*) are observed throughout Southern Scandinavia around 4000-3700 BC (Fig. 1). Recently presumed domesticated cows from Rosenhof LA 58 and 83 were dated to 4700 BC. But after a DNA-analysis, they turned out to be small aurochs (Hartz and Lübke 2005; Noe-Nygaard *et al.,* 2005). Another early cow tooth from Lollikhuse was dated to 5890±55 BP (4929-4612 cal BC. AAR-7410-2). It is probably an exotic pendant showing direct or indirect contacts with farming societies in Central Europe (Sørensen 2005, 305). Cow bones from Smakkerup Huse have also been used to promote the fact, that Ertebølle hunters had access to domesticated animals (Price and Gebauer 2006). These bones where dated to 5059±68 BP (3981-3701 cal BC. AAR-3316) and 5060±61 BP (3968-3711 cal BC. AAR-3317) and found in stratified Late Atlantic refuse layers. Unfortunately the actual site was eroded by transgressions and regressions in the Subboreal period, as such these bones could belong to an Early Funnel Beaker occupation. Sheep and goat (*Ovis/Capra*) also appear in Southern Scandinavia during the time 4000 to 3800 BC and a few centuries later in Western Sweden. Domesticated pigs (*Sus domesticus*) and wild boars (*Sus scrofa*), found together in a pit from the Early Neolithic site of Almhov in Scania, has been dated to 4960±50 BP (3937-3645 cal BC. Ua-22166) (Rudebeck 2010, 112ff). However, the identification of domesticated pigs has proven difficult and often judged on the fact, that they were smaller than wild boars and had different length and anterior breadth of the M3 (Magnell 2005; Brinch Petersen and Egeberg 2009, 562). Future DNA analysis could resolve this problem. Currently, there is no secure archaeological evidence of whole domesticated animals (except the dog) earlier than 4000 BC in Southern Scandinavia. Summarizing the early dates of cereal grains and domesticated animals clearly demonstrates that we are dealing with quick expansion (Hallgren 2008; Price and Noe-Nygaard 2009, 198ff; Sørensen 2012; Sørensen and Karg 2012).

Cultivation and crop processing

Plough marks found below a long barrow at Højensvej 7 near Egense on Funen covered an area of 85 square meters, thus illustrating an intensive cultivation (Beck 2009; in press). Some of the plough marks were cut by a pit, which was dated by a hazelnut shell to 4900±40 BP (3770-3637 cal BC. POZ-28068), thus proving a very early date of these plough marks. Currently they are the earliest from Southern Scandinavia. Other plough marks with more limited extensions have been documented below a few long barrows in Northern Germany, Jutland and Funen (Jørgensen 1977, 7; Fischer 1980, 23; Ebbesen 1992, 96; Mischka 2011, 742). We can conclude that cultivation of larger fields using the ard, in order to get maximum yield out of the soil, was present already from the beginning of the Early Neolithic.

Cereal crop processing and threshing waste from Emmer (*Triticum dicoccoides*) used as chaff tempering in clay discs has been found in some pits from the Early Neolithic site at Lisbjerg Skole near Århus (Skousen 2008, 124). 14C dates of hazelnut shells from the pits (A-2087, A-2092 and A2165) dates the material to 5190±90 BP (4251-3785 cal BC. AAR-8542) and to 4975±55 BP (3942-3651 cal BC. AAR-9225). Straw or chaff tempering is also found in clay discs from the Early Neolithic site of Store Valby (Becker 1954, 134; Helbæk 1954, 198; Nielsen 1984, 119). Evidence of crop processing is supported by the existence of quern stones. These have been reported from Early Neolithic sites in Denmark and Sweden (Erantisvej, Vallensgård I, Almhov, Fågelbacken and Skogsmossen), which also contain short necked funnel beakers (Staal 2005; Hallgren 2008, 211; Nielsen 2009, 14; Rudebeck 2010, 112). Moreover, wear on sickles from Early Neolithic sites also document harvesting activities (Juel Jensen 1994). The 14C dates of the pits at Lisbjerg Skole are very important, because they document cultivation and crop processing, which could be earlier than the Ulmus-fall, which has been wiggle-matched to 3870 BC (Andersen and Rasmussen 1993, 125). Arguably, pollen analyses are showing cultivation from 3600-3500 BC, which does not correspond with the evidences from the archaeological record. Most of the pollen records have been taken from smaller lakes or bogs, thus showing the environmental change on a very local scale. Cereal pollen is rarely detected in these pollen diagrams, because wheat and barley are self-pollinated species, which means that the pollen does nor spread over long distances (Diot 1992).

However, pollen found underneath long barrows demonstrations a higher rate of Ribwort plantain (*Plantago lanceolata*), which indicates clearances of the forest during the Early Neolithic (Odgaard 1994, 1; Rasmussen 2005, 1116; Sjögren 2006; Westphal 2009, 97). Maybe many pollen diagrams are showing initial stages of slash and burn cultivation combined with animal husbandry (Brinch Petersen and Egebjerg 2009, 560). The archaeological evidence is clearly proving the beginning of an agrarian subsistence strategy with cultivation from 4000 BC and onwards. The early appearance of a complete agrarian technology together with the evidences of a quick expansion could indicate that we are dealing with smaller groups of Central European pioneering farmers entering Southern Scandinavia around 4000 BC (Nielsen 1984, 116ff; Madsen 1987, 229ff; Kristiansen 1988, 27ff; Klassen 2004; Rowley-Conwy 2011, 431ff). Let us go through the secondary evidences from these Early Neolithic agrarian societies.

The secondary evidence

A different material culture, which occurs in Southern Scandinavia in the beginning of the 4th millennium BC points towards migrating farmers from Central Europe, which are expanding into Southern Scandinavia. It consists of pointed butted axes (Nielsen 1977, 65), jade axes (Klassen 2004), battle axes (Zápotocký 1992; Ebbesen 1998, 77), short necked funnel beakers (Koch 1998), clay discs (Davidsen 1974, 5) and copper artifacts (Klassen

2000). The structures includes two aisled houses (Nielsen 1997, 9), flint mines (Becker, 1980, 456; Olausson *et al.*, 1980, 183), long barrows (Rudebeck 2002, 119) and later on Sarup enclosures (Andersen 1997) and long dolmens (Ebbesen 2011).

Battle axes and pointed butted flint axes

The distribution of the battle axes are significant, because they illustrate dense concentrations in Central Europe, thus indicating an origin of these pioneering farmers within the Michelsberg and Baalberg cultures (Lüning 1968; Zápotocký 1992; Ebbesen 1998, 77; Hallgren 2008). The earliest type of battle axes, type 1 in Ebbesens typology (1998, 77) or type FI-III in Zápotockýs typology (1992), has been found in the Dragsholm burial. In this burial an antler pick was dated to 5090 ±65BP (4036-3712 cal BC AAR- 7418-2) and a human bone was dated to 5102±37 (3973-3797 cal BC. AAR-7416-2), thus dating the battle axes to the beginning of the 4th millennium BC (Brinch Petersen 2008, 33).

Particular important is also the distribution pattern of polished pointed butted axes, which is connected to Early Neolithic sites with short necked funnel beakers. Three types of pointed butted axes have been identified. Type 1 has an oval cross section, whereas type 2 and 3 has a three and four sided cross section (Nielsen 1977). Pointed butted flint axes from 14C dated contexts demonstrate overlap between the three types, which is also observed in the depositions of the axes. However type 1 has never been found together with type 3 or any of the thin butted axes (Rydbeck 1918, 9; Karsten 1994, 226; Rosenberg 2006), thus supporting the typology originally proposed by Nielsen (1977) (Fig. 3, 4). The distribution of the pointed butted axes illustrated a rather dense inland habitation during the Early Neolithic (Fig. 2). They seem to concentrate in regions with light easily arable soils. The pointed butted flint axes from Southern Norway are probably connected to the Early Neolithic agrarian expansion (Henningsmoen 1980; Østmo 1988; Prøsch-Danielsen 1996; Glørstad 2010, 275), whereas the ones from Trøndelag represent long exchange patterns with hunter-gatherer groups in the northern parts of Scandinavia. There must have been a huge and systematic production and distribution of these axes, which is revealed by clear concentrations of pointed butted axes near the flint mines on Stevns (now eroded by the sea) in Eastern Zealand and Södra Sallerup in Scania.

Flint mining and exchange system

Deep mining after flint is a characteristic feature in the Central European Michelsberg Culture (4400-3500 BC) (Lüning 1968). The mines at Spiennes in Southern Belgium, Rickholt in the Netherlands and Jablines/Le Haute Château in Northern France were all opened from 4400 to 4200 BC. (Bostyn and Lanchon 1992; Collet *et al.*, 2004, 151ff; Grooth *et al.*, 2011, 77). Preforms of pointed butted axes found in all these mines prove, that we are dealing with systemized production. The earliest evidence of mining in Southern Scandinavia is documented at Södra Sallerup and Hov, which has been dated to 4000 cal BC (Olausson *et al.*, 1980, 183; Personal communication Elisabeth Rudebeck and Jens Henrik Bech). The practice of mining flint probably came over together with these Central European pioneering farmers.

Reasons for the expansion

The reason for the expansion is still uncertain. One of the reasons could be the relatively easy access to some of the best flint resources in Northern Europe. Other possible explanations include growing population pressure combined with the fact, that cultivation by slash and burn method requires considerable space. These factors could have motivated some Central European farmers to cultivate new lands in the north. Recently Rowley-Conwy (2011) has suggested that pioneering farmers expanded to the north by leap-frog, punctuated or sporadic immigration (Moore 2001, 395ff). A similar model has been presented by Zilhao (2001, 14180) explaining a fast Neolithic expansion in the Mediterranean. The expansion towards Scandinavia happened so fast and covered such huge distances, that boats must have been used as indicated by very early Neolithic agrarian habitations on islands like Bornholm and Gotland (Lindqvist and Possnert 1997, 73; Casati and Sørensen 2006, 39; Nielsen 2009, 9).

Cultural dualism

During the Early Neolithic an agrarian way of life was practised on inland sites at the same time as hunting and fishing took place on sites near the coast, fjords or larger inland lakes. Are we dealing with commuting farmers or a cultural dualism? If hunter-gatherers went to the effort of keeping domesticated animals all year round, they would have to collect huge amounts of food for the winter months. In order to get enough food it would be necessary to cultivate larger fields of grain and grass for drying straw and hay for the winter. This would require a long term skill in agrarian technologies in order to succeed. If these hunter-gatherers were to succeed as farmers, they would gradually need to integrate with agrarian societies. Especially the use of slash and burn technique to clear the forest for cultivation required a necessity to plan several years ahead. Several experiments has shown, that after two to three years of cultivation the nourishment in the soil was used up with the consequence, that the yield would fall drastically (Lünning 2000, 174; Ehrmann *et al.*, 2009, 44; Schier 2009, 15). The field could hereafter be used as grazing areas for domesticated animals. But in order to continue cultivation it was necessary to start all over in another area, thus proving that this method requires access to huge areas. Recently, Kind (2010, 457) has proposed, that the transition towards agriculture is determined by an intensified social interaction between local hunter-gatherers and pioneering farmers, who is characterized as the "managers of neolithisation". The Dragsholm man, who was buried in a kitchen midden and

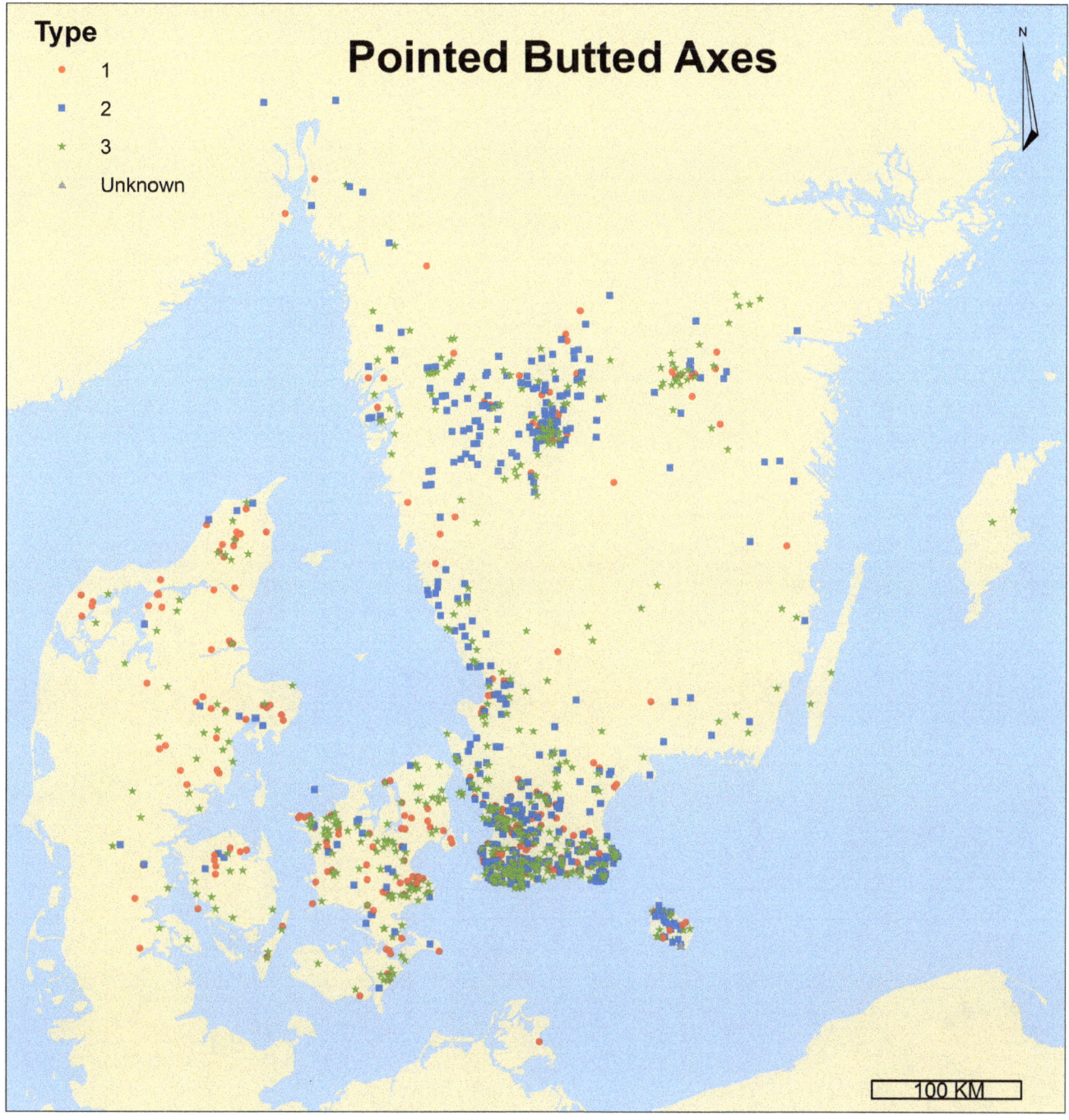

Fig. 2. Distribution of pointed butted axes in Southern Scandinavia based on data from Denmark (Brøndsted 1938, 130ff), Scania (Hernek 1988, 216ff), Bohuslän, Dalsland, Halland and Vester Götland (Blomqvist 1990), Bornholm (Nielsen 2009, 16ff), Middle parts of Sweden (Hallgren 2008), Southern Norway (Østmo 1986, 190ff), Middle parts of Norway (Valen 2007) and own studies.

equipped as a warrior, could be interpreted as one of these pioneering farmers expanding into a new territory with a complete agrarian technology and ideology, thus initiating the beginning of a cultural dualism in that area (Brinch Petersen 2008, 33ff).

Cultural dualism could be interpreted, when actual cultivation is found on a hunter-gatherer site. The Bjørnsholm kitchen midden could be one of these sites,

because the pollen of barley (*Hordeum*) and wheat (*Triticum*) was found under the neighboring long dolmen (Andersen and Johansen 1992, 38; Andersen 1993, 59). Visborg could be another example, because a burned layer under the kitchen midden indicates the use of the slash and burn clearance (Andersen 2008, 69). Another method of documenting a cultural dualism is by conducting DNA analysis. The burial site of Ostorf in Northern Germany was originally interpreted as a hunter-gatherer enclave

Depositions of pointed butted flint axes and their combination of types								
Site	*Region*	*Nr. of axes*	*Polished or unpolished*	*1*	*2*	*3*	*Thin*	*Refference*
Järavallen	Scania	11	unpolished	x				rydbeck 1918, 9
Hammelen	Scania	2	unpolished	x				rydbeck 1918, 9
Lackalänga	Scania	1+grinding stone	unpolished	x				karsten 1994, 226
Svedala	Scania	1+grinding stone	polished	x				rydebeck 1918, 9
Grönby	Scania	8	unpolished	x				nielsen 1977, 121
Arrie	Scania	4	unpolished	x	x			rydbeck 1918, 9ff
Ravnekær	Bornholm	5	polished and unpolished		x			P. O. Nielsen personal com.
Karaby	Scania	2	unpolished		x			Rydbeck 1918, 9
Dalby	Scania	2	polished		x			Rydbeck 1918, 12ff
Borgeby	Scania	2	polished		x			Rydbeck 1918, 12ff
V. Ågården	Vendsyssel	2	unpolished		x			Nielsen 1977, 121
Eslöv	Scania	2	unpolished		x			Nielsen 1977, 121
Fränninge	Scania	1+grinding stone	polished		x			Karsten 1994, 309
V. Ågården	Vendsyssel	3	unpolished		x	x		nielsen 1977, 121
Li Markie nr. 7	Scania	3	unpolished		x	x		rydbeck 1918, 11ff
Gualöv	Scania	3	polished		x	x	x	Karsten 1994, 348
Vanstad	Scania	2	polished			x		Rydbeck 1918, 16ff
Smeby Slöta	Västergötaland	5	polished			x		Nielsen 1977, 121
Ullerødgård	Zealand	3	polished			x	x	Rosenberg 2006
Kvistofta	Scania	3	polished			x	x	Karsten 1994, 215
Skegrie	Scania	2	unpolished			x	x	Karsten 1994, 294
Skurup	Scania	10	polished and unpolished			x	x	Karsten 1994, 303
Svedala	Scania	11	polished and unpolished			x	x	Karsten 1994, 274
Södra Åsum	Scania	2	polished			x	x	Karsten 1994, 310
Fjälkinge	Scania	2	polished and unpolished			x	x	Karsten 1994, 343
Kverrestad	Scania	3	polished			x	x	karsten 1994, 328
Öster Sönnarslöv	Scania	2	unpolished			x	x	Karsten 1994, 347
Hörby	Scania	6	polished			x	x	Karsten 1994, 238
Bodarp	Scania	6	unpolished			x	x	karsten 1994, 282

Fig. 3. Table showing the depositions of pointed butted flint axes and their combination of types.

surrounded by agrarian societies, because the individuals had a high intake of aquatic resources (Lübke *et al.*, 2009, 130; Shulting 2011, 21). However, three burials contained Palaeolithic/Mesolithic haplogroups U5 and U5a, while four other burials contained Neolithic haplogroups J, K and T2e (Bramanti *et al.*, 2009, 139). The individuals at Ostorf illustrate a rare example of hunter-gatherers and possible farmers, which may have integrated with each other. Currently the archaeogenetic evidence is pointing in two directions. Skoglund and his team (Skoglund *et al.*, 2012) argue for a migration of farmers, whereas Melchior and his team (Melchior *et al.*, 2010) suggest that this is unlikely, and that there occurs an abrupt replacement of the Mesolithic hunter-gatherer population with a new Neolithic population in Southern Scandinavia.

Future research areas

The hypothesis regarding cultural dualism can be proved or disproved by C-14 dates, stable isotope and DNA analysis. If the DNA influx from pioneering farmers consists of Paleolithic/Mesolithic haplogroups (D and U4 and U5) representing Central European hunter-gatherers who became farmers, then it will be difficult to detect any differences (Bramanti *et al.*, 2009, 137). Stable isotope analysis has already been made by Tauber (1981) and Fischer *et al.*, (2007, 2125). Currently marine values of δ13C and δ15N are all extracted from Mesolithic hunter-gatherers, while most of the Early Neolithic samples, except one from Sejrø, are showing non marine values. It is clear that human bones from the Early Neolithic coastal/kitchen

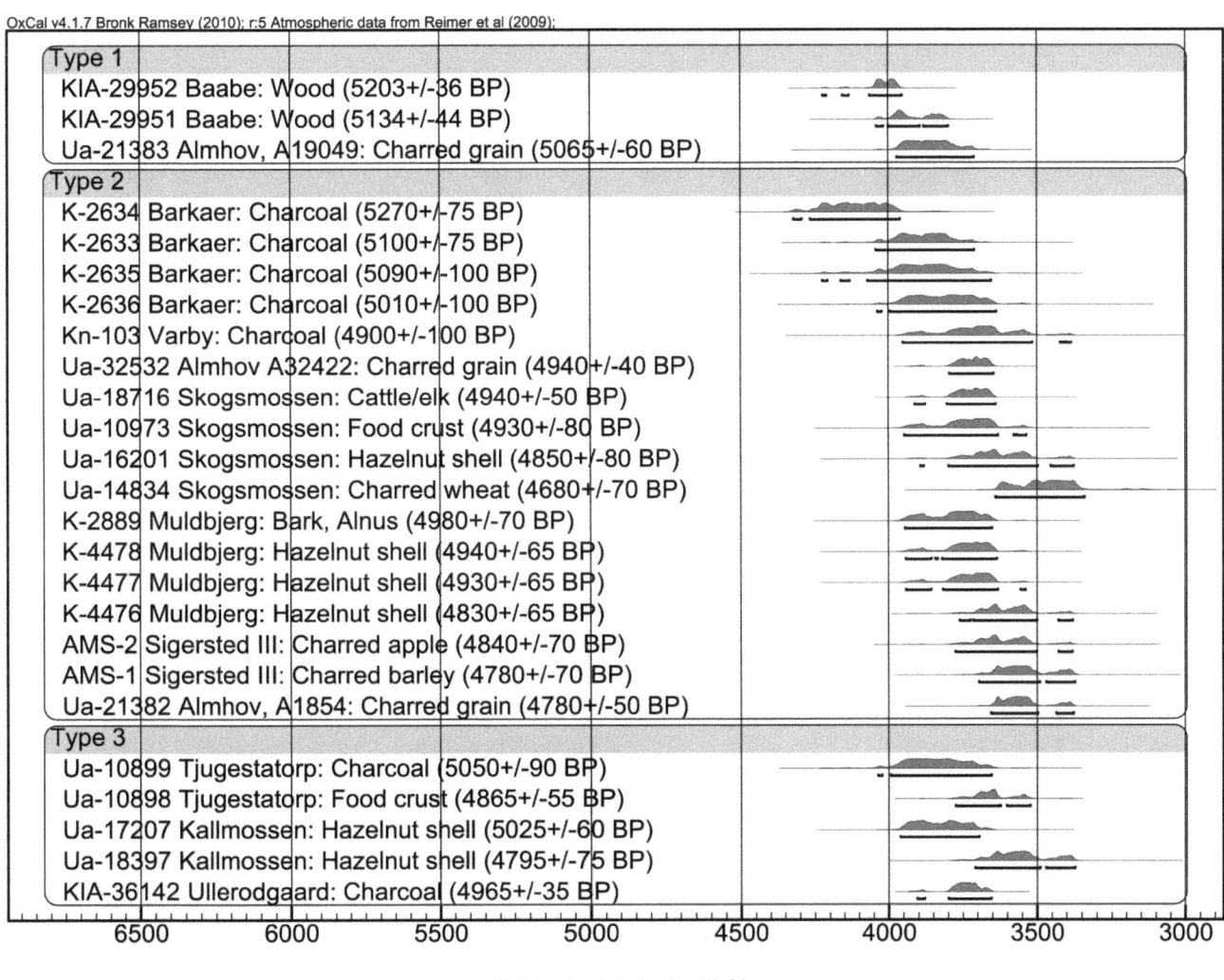

OxCal v4.1.7 Bronk Ramsey (2010); r:5 Atmospheric data from Reimer et al (2009);

Calibrated date (calBC)

Fig. 4. C-14 dates of Early Neolithic sites or contexts containing pointed butted axes. All radiocarbon dates have been calibrated using the OxCal v4.1.7 program. Data after: (Hirsch et al. 2008, 25ff; Rudebeck 2010, 112f; Liversage 1992, 59; Salomonsson 1970, 72; Hallgren 2008, 233ff; Troels-Smith 1957; Stafford 1999, 91; Koch 1998; Esben Aasleff personal communication)

midden sites are lacking in these analyses. The abrupt shift in δ13C values could be interpreted as a deliberately deselecting of marine food as a resource (Andersen *et al.*, 1986; Brinch Petersen and Egeberg 2009, 563; Milner *et al.*, 2004, 9). However, this does not necessarily mean that farmers moved away from the coastal areas, as funnel beaker hunting stations are documented through the Early and Middle Neolithic (Skaarup 1972). But it could reflect a gradual decline of marine resources.

Concluding remarks

The agrarian expansion during the Early Neolithic in Southern Scandinavia was a quick process lasting only a few centuries, between 4000 and 3700 BC. The speed of the expansion occurred so fast, that it must have involved smaller groups of migrating pioneering farmers possibly using boats as means of transportation. They originated from Middle Neolithic societies in Central Europe. The reason for the expansion is still uncertain, but growing population pressure combined with the fact, that cultivation

by slash and burn method requires substantial space, could motivate some pioneering farmers to move north. They brought with them a complete agrarian technology, structures, new material culture and ideology. The question of what happened to the local hunter-gatherers is still open for discussion. Either they became farmers really quickly within one or two generations and participated actively in the spreading of agrarian way of life. This could explain the settling of both inland and coastal sites, where they could exploit both agrarian resources together with hunting and fishing activities. Or we could be dealing with a cultural dualism consisting of pioneering farmers on inland oriented sites and hunter-gatherers on coastal sites. In this hypothesis the local hunter-gatherers quickly adopted the new material culture and domesticated animals, but in many cases they continued their hunter-gatherer lifestyle. Cultural dualism could be interpreted, when evidences of cultivation has been found on a hunter-gatherer site. Both explanations are plausible, but currently the archaeological evidences in Southern Scandinavia tend to favor a cultural dualism during the earliest part of the Early Neolithic. The transition towards an agrarian way of life in Scandinavia can therefore be interpreted as a much more complex and continuous process of migration, integration and gradual assimilation between neighboring farmers and hunter-gatherers happening at different speed from region to region depending on environmental, social as well as ideological factors.

References

Andersen, N. H. 1997. *The Sarup enclosures. The Funnel Beaker Culture of the Sarup site including two causewayed camps compared to the contemporary settlements in the area and other European enclosures.* Århus, Jutland Archaeological Society Publications XXXIII:1.

Andersen, S. H. 2008. The Mesolithic – Neolithic transition in Western Denmark seen from a kitchen midden perspective. A survey. In: H. Fokkens,. B. J. Coles., A. L. Van Gijn., J. P. Kleijne., H. H. Ponjee and C. G. Slappendel (eds.), *Between Foraging and Framing*, 67-74. Leiden, Analecta Prehistorica Leidensia.

Andersen, S. H. and Johansen, E. 1992. An Early Neolithic Grave at Bjørnsholm, North Jutland. *Journal of Danish Archaeology* 9, 38-58.

Andersen, S. H., Constandse Westermann, T., Newell, R. R., Gillespie, R., Gowlet, J. A. J. and Hedges, R. E. M. 1986. New radiocarbon results from two Mesolithic burials in Denmark. In: J. A. J. Gowlet and R. E. M. Hedges (eds.), *Archaeological results from Accelerator Dating*, 39-43. *Monograph 11. Oxford University Committee for Archaeology.*

Andersen, S. Th. 1993. Early and Middle Neolithic agriculture in Denmark: pollen spectra from soils in burial mounds of the Funnel Beaker Culture. *Journal of European Archaeology* 1, 153-180.

Andersen, S. Th. and Rasmussen, K. L. 1993. Radiocarbon wiggle-dating of elm declines in north-west Denmark and their significance. *Vegetation History and Archaeobotany* 2, 125-135.

Beck, M. R. 2009. Lå Danmarks første pløjemark ved Egense? *Svendborgs Museums* Årbog 2009, 7-16.

Beck, M. R. in press. Højensvej høj 7 – en tidligneolitisk langhøj med flere faser ved Egense, Svendborg. To be published in: *Aarbøger for Nordisk Oldkyndighed og Historie.*

Becker, C. J. 1954. Stenalderbebyggelsen ved Store Valby i Vestsjælland. Problemer omkring tragtbægerkulturens ældste og yngste fase. *Aarbøger for Nordisk Oldkyndighed og Historie* 1954, 127-197.

Becker, C. J. 1980. Dänemark. In: G. Weisberger (ed.), *5000 Jahre Feuersteinbergbau. Die Such nach dem Stahl der Steinzeit. Ausstellung im Deutschen Bergbau-Museum Bochum vom 24 Oktober 1980 bis 31. Januar 1981*, 456-471. Bochum. Deutschen Bergbau-Museum.

Blomqvist, L. 1990. *Neolitisk Atlas över Västra Götland.* Falköping. Norders Bokhandel.

Bostyn, F. and Lanchon, Y. 1992. *Jablines: Le Haut Château (Seine-et-Marne), Un miniére de silex au Néolithique.* Paris, Documents d'Archéologie Francaise.

Bramanti, B., Thomas, M. G., Haak, W., Unterlaender, M., Jores, P., Tambets, K., Antanaitis-Jacobs., M. N. Haidle., Jankauskas, R., Kind, C. J., Lueth, F., Terberger, T., Hiller, J., Matsumura, S., Forster, P. and Burger, J. 2009. Genetic discontinuity between local hunter-gatherers and central Europe's first farmers. *Science* 326, 137-140.

Brinch Petersen, E. 2008. Warriors of the Neolithic TRB-Culture. In: Z. Sulgostowska and A. J. Tomaszewski (eds.). *Man Millenia Environment. Studies in honour of Roman Schild*, 33-38. Warsaw. Polish Academy of Science.

Brinch Petersen, E. and Egeberg, T. 2009. Between Dragsholm I and II. In: L. Larsson, F. Lüth and T. Terberger (eds.). *Innovation and Continuity – Non Megalithic Mortuary Practices in the Baltic. New methods and research into the development of Stone Age Society*, vol 88, 447-467. Bericht der Römisch-Germanischen Kommission. Frankfurt. Philipp von Zabern.

Brønsted, J. 1938. *Danmarks Oldtid I.* First edition. København. Gyldendal.

Casati, C. and Sørensen, L. 2006. Bornholm i ældre stenalder. Status over kulturel udvikling og kontakter. *Kuml* 2006, 9-58.

Collet, H., Collette, O. and Woodbury, M. 2004. Indices d'extraction et de taille du silex datant du Néolithique recent dans la Carriére CBR à Harmignies. Note préliminaire. *Notae Praehistoricae* 24, 151-158.

Davidsen, K. 1974. Neolitiske lerskiver belyst af danske fund. *Aarbøger for Nordisk Oldkyndighed og Historie* 1973, 5-72.

Diot, M.-F., 1992. Études palynologiques de blés sauvages et domestiques issus de cultures expérimentales. In: P. C. Anderson (ed.), *Préhistoire de l'agriculture: nouvelles approaches expérimentales et ethnographiques*, 107-111. Monographie du CRA n° 6. CNRS, Paris.

Ebbesen, K. 1992. Simple, tidligneolitiske grave. *Aarbøger for Nordisk Oldkyndighed og Historie* 1992, 47-102.

Ebbesen, K. 1998. Frühneolithische Streitäxte. *Acta Archaeologica* 69, 77-112.

Ebbesen, K. 2011. *Danmarks Megalitgrave.* København. Forfatterforlaget Attika.

Ehrmann, O., Rösch, M. and Schier, W. 2009. Experimentelle Rekonstruktion eines jungneolitischen Wald-Feldbaus mit Feuereinsatz – ein multidisziplinäres Forschungsprojekt zur Wirtschaftsarchäologie und Landschaftsökologie. *Prähistorische Zeitschrift* Band 84, 44-72.

Fischer, A. 2002. Food for Feasting ? An evaluation of explanations of the neolithisation of Denmark and southern Sweden. In: A. Fischer and K. Kristiansen (eds.), *The neolithisation of Denmark - 150 years of debate*, 341-393. Sheffield. J. R. Collis Publications.

Fischer, A., Olsen, J., Richards, M., Heinemeier, J., Sveinbjörnsdottir, A. E. and Bennike, P. 2007. Coast-inland mobility and diet in the Danish Mesolithic and Neolithic: evidence from stable isotope values of humans and dogs. *Journal of Archaeological Science* 34, 2125-2150.

Fischer, C. 1980. Brendkroggården. En langhøj/langdysse ved Salten i Midtjylland. *Antikvariske Studier* 4, 23-30.

Glørstad, H. 2010. *The Structure and History of the Late Mesolithic Societies in the Oslo fjord area 6300-3800 BC*. Lindome. Bricoleur Press.

Grooth, M. E. Th., Lauwerier, R. C. G. M. and Schegget, M. E. 2011. New C-14 dates from the Neolithic flint mines at Rijckholt-St. Geertruid, the Netherlands. In. M. Capote, S. Consuegra, P. Diaz del Rio and X. Terradas (eds.), *Proceedings of the 2nd International Conference of the UISPP Commission on Flint Mining in Pre- and Protohistoric Times (Madrid, 14-17 October 2009)*, 77-89. BAR International Series 2260. Oxford.

Hadevik, C. 2009. Trattbägerkulturen i Malmöområdet. En sammenställing med fokus på byggnader, gravar och rituelle gropar. In. C. Hadevik and M. Steineke (eds.), *Spåren i marken – tematisk rapportering från Citytunnelprojektet*, 13-90. Rapport nr. 48. Malmø. Malmø Museer.

Hallgren, F. 2008. *Identitet i Praktik. Lokala, regionala och överregionala sociala sammanhang inom nordlig trattbägerkultur.* Uppsala.

Hartz, S and Lübke, H. 2005. Zur chronostratigraphischen Gliederung der Ertebølle-Kultur und frühesten Trichterbecherkultur in der südlichen Mecklenburger Bucht. *Bodendenkmalpflege in Mecklenburg-Vorpommeren. Jahrbuch* 2004, vol. 52, 119-143. Lübsdorf.

Heinemeier, J. and Rud, N. 2000. Danske arkæologiske AMS C-14 dateringer, Århus 1999. *Arkæologiske Udgravninger i Danmark* 2000, 296-313.

Heinemeier, J. 2002. AMS C-14 dateringer, Århus 2001. *Arkæologiske udgravninger i Danmark* 2001, 263-292.

Heinemeier, J. and Rud, N. 1998. Danske arkæologiske AMS C-14 dateringer, Århus 1997. *Arkæologiske Udgravninger i Danmark* 1997, 282-292.

Heinemeier, J. and Rud, N. 1999. Danske arkæologiske AMS C-14 dateringer, Århus 1998. *Arkæologiske Udgravninger i Danmark* 1998, 327-345.

Helbæk, H. 1954. Kornavl i Store Valby. *Aabøger for Nordisk Oldkyndighed og Historie* 1954, 198-204.

Henningsmoen, K. 1980. Trekk fra floraen i Vestfold. In: V. Møller (ed.), *Bygd og by i Norge, Vestfold*, 163-175. Oslo. Gyldendal Norsk Forlag.

Hernek, R. 1988. Den spetsnackiga yxan av flinta. *Fornvännen* 83, 216-223.

Hirsch, K., Klooss, S. and Klooss, R. 2007. Der endmesolithisch-neolithische Küstensiedlungsplatz bei Baabe im Südosten der Insel Rügen. *Bodendenkmalpflege in Mecklenburg- Vorpommeren. Jahrbuch* 2007, vol. 55, 11-51. Lübsdorf.

Jennbert, K. 1984. *Den Produktiva Gåvan. Tradition och innovation i Sydskandinavien för omkring 5300 år sedan.* Acta Archaeologica Lundensia, Serie 4⁰ 16. Bonn. Rudolf Habelt Verlag.

Johansson, G., Westergaard, B., Artelius, T and Nieminen, J. 2011. *Arkeologiska undersökningar. Boplatser och gravar vid Viskan i Veddige. Fem fornlämningar undersökta för riksväg 41. Halland, Veddige socken,* Vabränna 1:5, 10:5, 3:39 och 3:33; Kullagård 1:13 och Järlöv 7:3, RAÄ 323, 322, 128b, 320 och 321. Riksantikvrieämbetet UV Rapport 2011:26. Riksantikvrieämbetet, Sweden.

Juel Jensen, H. 1994. *Flint tools and plant working – hidden traces of stone age technology*. Århus. Århus University Press.

Jørgensen, E. 1977. Brændende dysser. *Skalk*, vol. 5, 7-13.

Karg, S. and Harild, J. 2009. Makrofossilanalyse af gytjeprøver fra Ronæs Skov. In: S. H. Andersen (ed.), *Ronæs Skov. marinarkæologiske undersøgelser af en kystboplads fra Ertebølletid*, 231-233. Århus. Jysk Arkæologisk Selskabs skrifter.

Karsten, P. 1994. *Att kasta yxan i sjön*. Acta Archaeologica Lundensia, Series in 8⁰, No. 23. Stockholm. Almqvist and Wiksell International.

Kind, C-J. 2010. Diversity at the transition – a view from the Mesolithic, In: D. Gronenborn and J. Petrasch (ed.), *Die Neolithisierung Mitteleuropas – The Spread of the Neolithic to Central Europe, International Symposium, Mainz 24 June – 26 June 2005*, 449-460. Mainz. Römisch-Germanisches Zentralmuseums.

Klassen, L. 2000. *Frühes Kupfer im Norden Untersuchungen zur Chronologie, Herkunft und Bedeutung der Kupferfunde der Nordgruppe der Trichterbecherkultur.* Århus. Jysk Arkæologisk Selskabs skrifter.

Klassen, L. 2004. *Jade und Kupfer Untersuchungen zum Neolithisierungsprozess im westlichen Ostseeraum unter besonderer Berücksichtigung der Kulturentwicklung Europas 5500-3500 BC*. Århus. Jysk Arkæologisk Selskabs skrifter.

Koch, E. 1987. Ertebølle and Funnel Beaker Pots as Tools. *Acta Archaeologica*, vol. 57, 107-120.

Koch, E. 1998. *Neolithic Bog Pots from Zealand, Møn, Lolland and Falster.* Nordiske Fortidsminder serie B, vol. 16. København. Det Kongelige Nordiske Oldskriftselskab.

Kristiansen, K. 1988. Det tidligste agerbrug i Danmark (4000-3600 f.Kr.). In: C. Bjørn, T. Dahlerup, S. P. Jensen and E. Helmer Pedersen (eds.), *Det danske landbrugs historie bind 1, oldtid og middelalder*, 21-40. Odense. Landbohistorisk Selskab.

Larsson, M. 1992. The early and middle Neolithic Funnel Beaker Culture in the Ystad area (Southern Scania). economic and social change, 3100-2300 BC. In: L. Larsson, J. Callmer and B. Stjernquist (eds.), *The Archaeology of the Cultural Landscape*, 17-90. Acta Archaeologica Lundensia 4 (19). Stockholm. Almqvist and Wiksell International.

Lindqvist, C. and Possnert, G. 1997. The subsistence economy and diet at Jakobs/Ajvide and Stora Förvar, Eksta parish and other prehistoric dwelling and burial sites on Gotland in long-term perspective. In: G. Burenhult (ed.). *Remote Sensing Vol. I. Theses and papers in North-European archaeology* 13:a, 29-90. Department of archaeology, Stockholm University. Hässleholm.

Liversage, D. 1992. *Barkær, Longbarrows and Settlements.* Arkæologiske Studier vol. IX. København. Akademisk Forlag.

Lübke, H. Lüth, F. and Terberger, T. 2009. Fishers or farmers? The archaeology of the Ostorf cemetery and related Neolithic finds in the light of new data. In: L. Larsson, F. Lüth, T. Terberger (eds.), *Innovation and Continuity Non-Megalithic Mortuary Practices in the Baltic. New Methods and Research into the Development of Stone Age Society. International Workshop Schwerin 24-26 March 2006*. Bericht der Römisch-Germanischen Kommission, vol. 88, 307-38. Mainz. Philipp von Zabern.

Lüning, J. 1968. *Die Michelsberger Kultur. Ihre Funde in zeitlicher und räumlicher Gliederung.*

Bericht der Römisch-Germanischen Kommission, vol. 48, 1-350.

Lüning, J. 2000. *Steinzeitliche Bauern in Deutschland. Die Landwirtschaft im Neolithikum*. Universitätsforschungen zur prähistorischen Archäologie, Band 58. Bonn. Rudolf Habelt.

Madsen, T. 1987. Where did all the Hunters go? An assessment of an epoch-making episode in Danish Prehistory. *Journal of Danish Archaeology* 5, 229-239.

Magnell, O. 2005. *Tracking wild boar and hunters. Osteology of wild boar in Mesolithic South Scandinavia.* Studies in Osteology I. Acta Archaeologica Lundensia Series in 8° No 51. Almqvist and Wiksell International, Stockholm.

Melchior, L., Lynnerup, N., Siegismund, H. R., Kivisild, T., Dissing, J., 2010. Genetic Diversity among Ancient Nordic Populations. *PLoS ONE*, July, Vol. 5, Issue 7, 1-9.

Milner, N., Craig, O. E., Bailey, G. N., Pedersen, K. and Andersen, S. H. 2004. Something fishy in the Neolithic? A re-evaluation of stable isotope analysis of the Mesolithic coastal populations. *Antiquity* 78, 9-22.

Mischka, D. 2011. The Neolithic burial sequence at Flintbek LA 3, north Germany, and its cart tracks: a precise chronology. *Antiquity* 85, 742-758.

Moore, J. H. 2001. Evaluating Five Models of Human Colonization. *American Anthropologist* 103, nr. 2, 395-408.

Møller, P. F., Odgaard, B., Rasmussen, P. and Aaby, B. 2010. Urskovslandskabets naturlige åbenhed. *Skoven* nr. 10, 450-453.

Nielsen, A. B. and Buchwald, E. 2010. Urskovslandskabets åbenhed og græsningens betydning. *Skoven* nr. 10, 88-93.

Nielsen, P. O. 1977. Die Flintbeile der Frühen Trichterbecherkultur in Dänemark. *Acta Archaeologica* 48, 61-138.

Nielsen 1984. P. O. Nielsen, De første bønder. Nye fund fra den tidligste Tragtbægerkultur ved Sigersted. *Aarbøger for Nordisk Oldkyndighed og Historie* 1984, 96-126.

Nielsen, P. O. 1997. De ældste langhus. Fra toskibede til treskibede huse i Norden. *Bebyggelseshistorisk tidskrift* nr. 33, 9-30.

Nielsen, P. O. 2009. Den tidligneolitiske bosættelse på Bornholm. In. A. Schülke (ed.), *Plads og Rum i Tragtbægerkulturen*, 9-24. Nordiske Fortidsminder Serie C. København. Det Kongelige Nordiske Oldskriftselskab.

Noe-Nygaard, N., Price, T. D. and Hede, S. U. 2005. Diet of aurochs and early cattle in southern Scandinavia: evidence from N15 and C-13 stable isotopes. *Journal of Archaeological Science* 32, 855-871.

Odgaard, B. 1994. The Holocene vegetation history of northern West Jutland, Denmark. *Opera Botanica*, 123, 1-171.

Olausson, D., Rudebeck, E. and Säfvestad, U. 1980. Die südschwedischen Feuersteingruben – ergebnisse und Probleme. In: G. Weisberger (ed.), *5000 Jahre Feuersteinbergbau. Die Such nach dem Stahl der Steinzeit. Ausstellung im Deutschen Bergbau-Museum Bochum vom 24 Oktober 1980 bis 31. Januar 1981*, 183-204. Bochum. Deutschen Bergbau-Museum.

Pelgar, S. M. and Birks, H. J. B. 1993. The mid-Holocene Ulmus fall at Diss Mere, South-East England – disease and human impact? *Vegetation History and Archaeobotany* 2, 61-68.

Persson, P. 1999. *Neolitikums början. Undersökningar kring jordbrukets introduktion i Nordeuropa.* Göteborg. University of Gothenburg and University of Uppsala.

Petersen, P. V. 1993. *Flint fra Danmarks Oldtid*. København. Høst og Søn.

Pines, I. and Westwood, R. 2008. A Mark-recapture Technique for the Dutch Elm Disease Vector the Native Elm Bark Beetle, Hylurgopinus rufipes (Coleoptera:Scolytidae). *Arboriculture and Urban Forestry*, 34 (2), 116-122.

Price, T. D. and Gebauer, A. B. 2006. *Smakkerup Huse: a Late Mesolithic Coastal Site in Northwest Zealand, Denmark.* Århus. Århus University Press.

Price, T. D., Noe-Nygaard, N. 2009. Early Domestic Cattle in Southern Scandinavia. In: N. Finlay, S. McCartan, N. Milner and C. Wickham-Jones (eds.), *From Bann Flakes to Bushmills*, 198-210. Oxford, Oxbow Press.

Prøsch-Danielsen, L. 1996. Vegetation history and human impact during the last 11500 years at Lista, the southernmost part of Norway: based primarily on professor Ulf Hafstens material and diary from 1955-1957. *Norsk geografisk tidsskrift* 50/2, 85-99.

Rasmussen, P. 2005. Mid- to late Holocene land-use change and lake development at Dallund Sø, Denmark: vegetation and land-use history inferred from pollen data. *The Holocene* 15, 2005, 1116-1129.

Reimer, P.J., Baillie, M.G.L., Bard, E., Bayliss, A., Beck, J.W., Blackwell, P.G., Bronk Ramsey, C., Buck, C.E., Burr, G.S., Edwards, R.L., Friedrich, M., Grootes, P.M., Guilderson, T.P., Hajdas, I., Heaton, T.J., Hogg, A.G., Hughen, K.A., Kaiser, K.F., Kromer, B., McCormac, F.G., Manning, S.W., Reimer, R.W., Richards, D.A., Southon, J.R., Talamo, S., Turney, C.S.M., van der Plicht, J. and Weyhenmeyer, C. E. 2009. IntCal09 and Marine09 Radiocarbon Age Calibration Curves, 0–50,000 Years cal BP. *Radiocarbon*, 51(4), 1111-1150.

Rosenberg, A. 2006. *Beretning for Ullerødgård. Beretning over den arkæologiske forundersøgelse og udgravning på matr. 3x m.fl. Ullerød By, Tjæreby. (NFHA 2424).* Excavation rapport from Folkemuseet in Hillerød. http://www.folkemuseet.dk/Bygherre/A2424_net.pdf

Rowley-Conwy, P. 2011. Westward Ho! The spread of agriculture from Central Europe to the Atlantic. *Current Anthropology* 52, 431-451.

Rudebeck, E. 2002. Likt och olikt i de sydskandinaviska långhögarna. In: L. Larsson (ed.), *Monumentala gravformer i det äldsta bondesamhället*, 119-146. Report Series No. 83. University of Lund. Department of Archaeology and Ancient History.

Rudebeck, E. 2010. I trästoderna skugga - monumentala möten i neolitiseringens tid. In: B. Nilsson, and E. Rudebeck (eds.), *Arkeologiska och förhistoriska världar. Fält, erfarenheter och stenåldersplatser i sydvästra Skåne*, 83-251. Malmö Museer, Arkeologienheten. Malmö.

Ryberg, E. 2006. *Arkeologisk undersökning. Vägen till Veddige. Huslämningar och annat från neolitikum och skiftet bronsålder/järnålder. Halland, Veddige socken, Veddige 33:3, RAÄ 258.* UV Väst Rapport 2006:5. Riksantikvrieämbetet, Sweden.

Rydbeck, O. 1918. *Slutna mark- och mossfynd från stenåldern.* Lund. Lunds Universitets Historiska Museum.

Skaarup, J. 1972. *Hesselø – Sølager. Jagdstationen der südskandinavischen Trichterbecherkultur.* Arkæologiske Studier vol. 1. København. Akademisk Forlag.

Salomonsson, B. 1970. Die Värby-Funde. Ein Beitrag zur kenntnis der ältesten Trichterbecherkultur in Schonen. *Acta Archaeologica*, vol. 41, 55-95.

Schier, W. 2009. Extensiver Brandfeldbau und die Ausbreitung der neolitischen Wirtschaftsweise in Mitteleuropa und Südskandinavien am Ende des 5. Jahrtausends v. Chr. *Prähistorische Zeitschrift*, Band 84, 15-43.

Schulting, R. 2011. Mesolithic-Neolithic Transitions: An Isotopic Tour through Europe. In: R. Pinhasi and J.T. Stock (eds.), *Human Bioarchaeology of the Transition to Agriculture*, 17-41. New York. John Wiley and Sons.

Sjögren, K. G. 2006. *Ecology and Economy in Stone and Bronze Age Scania. Skånska spår – arkeologi längs Västkustbanan.* Riksantikvarieämbetet, UV Syd.

Skoglund, P., Malmström, H., Raghavan, M., Storå, J., Hall, P., Willerslev, E., Gilbert, T. P., Götherström, A., Jacobsson, M. 2012. Origins and Genetic Legacy of Neolithic Farmers and Hunter-Gatherers in Europe. *Science* 336, 466-469.

Skousen, H. 2008. *Arkæologi i lange baner. Undersøgelser forud for anlæggelsen af motorvejen nord om Århus.* Højbjerg. Moesgård Museum.

Staal, B. 2005. *Udgravningsberetning for den arkæologiske undersøgelse i forlængelse af arkæologisk forundersøgelse på Erantisvej i Nørre Alslev.* Nykøbing Falster. Museum Lolland-Falster.

Stafford, M. 1999. *From Forager to Farmer in Flint: a lithic analysis of the prehistoric transition to agriculture in southern Scandinavia.* Århus. Århus University Press.

Svensson, M. 2010. What time is it? *In Situ* 2009-2010, 7-26.

Sørensen, L. 2012. Pioneering farmers cultivating new lands in the North – The expansion of agrarian societies during the Neolithic and Bronze Age in Scandinavia. In: H. C. Gulløv, P. A. Toft and C. P. Hansgaard (eds.), *Challenges and solutions*, 87-124. Northern Worlds report from workshop 2 at the National Museum, 1. November 2011. Copenhagen. The National Museum.

Sørensen, L and Karg, S. 2012. The expansion of agrarian societies towards the North – new evidence for agriculture during the Mesolithic/Neolithic transition in Southern Scandinavia. To be published in: *Journal of Archaeological Science.*

Sørensen, S. A. 2005. Fra Jæger til bonde. In: C. Bunte., B. E. Berglund and L. Larsson (eds.). *Arkeologi och Natirvetenskap. Gyldenstiernska Krabberup Stiftelsens Symposium Nr. 6*, 298-309. Gyldenstiernska Krabberup Stiftelsen. Lund.

Tauber, H. 1981. C-13 evidence for dietary habits of prehistoric man in Denmark. *Nature* 292, 332-333.

Troels-Smith, J. 1957. Muldbjerg-bopladsen. Som den så ud for 4500 år siden. De første spor af agerbrug i Danmark. *Naturens Verden*, juli 1957, 1-30.

Valen, C. R. 2007. *Jordbruksimpulser i neolitikum og bronsealder i Nord-Norge? En revisjon av det arkeologiske gjenstandsmaterialet og de naturvidenskapelige undersøkelsene.* Unpublished MA thesis in archaeology. Tromsø. University of Tromsø.

Westergaard, B. 2008. *Arkeologisk undersökning. Trattbägare i O-bygd. Arkeologiska undersökningar längs E6 i Bohuslän, delen Lugnet-Skee. Bohuslän, Skee socken, Neanberg 1:14 och S:a Slön 2:4, Skee 1616.* UV Väst Rapport 2008:40. UV Väst. Riksantikvarieämbetet, Sweden.

Zápotocký, M. 1992. *Streitäxte des mitteleuropäischen Äneolithikums.* Weinheim, Acta Humaniora.

Zilhao, J. 2001. Radiocarbon evidence for maritime pioneer colonization at the origins of farming in west Mediterranean Europe. *Proceedings of the National Academy of Science*, November 20, Vol. 98, No. 24, 14180-14185.

Østmo, E. 1986. New observations on the funnel beaker culture in Norway. *Acta Archaeologica* vol. 55, 190-198.

Østmo, E. 1988. *Etableringen av jordbrukskultur i Østfold i steinalderen.* Universitetets Oldsaksamlings Skrifter, Ny rekke Nr. 10. Oslo. Universitetets Oldsaksamling.

Personal communication

Esben Aasleff. Folkemuseet in Hillerød.

Jens Henrik Bech. Thisted Museum.

Elisabeth Rudebeck. Sydsvensk Arkeologi AB.

Karl Göran Sjögren. University of Gothenburg.

CHAPTER 3

BEYOND THE NEOLITHIC TRANSITION - THE 'DE-NEOLITHISATION' OF SOUTH SCANDINAVIA

Rune Iversen

Abstract: *In South Scandinavia, the Funnel Beaker culture is synonymous with the emergence of Neolithic societies (c 4000 BC), the construction of megalithic monuments and agricultural lifestyle. After c 1300 years of existence the Funnel Beaker culture ceased and a culturally blurred period began.*

In the south-western parts of the Jutland Peninsula, the Single Grave culture emerged (c 2850 BC) expressing a high degree of cultural uniformity. In Eastern Denmark this uniformity was absent and instead the material culture shows a mixture of late Funnel Beaker, Pitted Ware and Single Grave culture elements. The question is whether the end of the Funnel Beaker culture in Eastern Denmark marks a period of decline and fragmentation or one of continuity and incorporation of new cultural elements and subsistence strategies. In particular the revival of hunter-fisher-gatherer strategies applied by the Pitted Ware culture represents a different economic focus than that held by the Funnel Beaker culture. The renewed focus on hunter-fisher-gatherer strategies, 1000 years after the introduction of agriculture, challenges the prevalent understanding of the dynamics behind the Neolithisation.

Keywords: *'de-Neolithisation', Neolithic lifestyle, mobile strategies, megalithic tombs, TRB, Single Grave culture, Pitted Ware culture, South Scandinavia, third millennium BC.*

In South Scandinavia, the Funnel Beaker culture is synonymous with the emergence of Neolithic societies, the construction of monumental tombs and agricultural life. Megalithic tombs and farming continued as essential features throughout the Neolithic (c 4000-1700 BC) but profound changes occurred with the emergence of Pitted Ware and Single Grave (Corded Ware) communities at the beginning of the third millennium BC. By analysing changes in the use of megalithic tombs, the importance of farming and settlement organisation, this paper will focus attention on the fact that the Neolithisation was not a unilinear process ending with the consolidation of farming and nucleated settlements. Instead, Neolithic life was changeable and shows the reintroduction of mixed subsistence strategies and mobile settlements at the beginning of the third millennium BC.

The construction and use of monumental tombs

Monumental tombs stand as iconic features of the Neolithic. The term comprises earthen long barrows and megalithic tombs, the latter subjected to a diversified reuse practice. From the beginning of the Early Neolithic (EN I/A and B) earthen long barrows were constructed in Denmark (Liversage 1992, 94-6; Thomas 1999, 131). This initial period of monument building was followed by the construction of dolmens starting c 3500 BC (EN II/C) and passage graves c 3300 BC equivalent to the beginning of the Middle Neolithic (MN I), cf. fig. 1.

In Denmark, estimated 25,000 megalithic tombs were built within a period of c 400 years. From antiquarian recordings about 7000 tombs are known (fig. 2) of which 2400 are preserved and protected today. Of the 2400 protected monuments, 700 are passage graves (Dehn *et al.* 2000, 7; Jensen 2001, 366-367; Dehn and Hansen 2006). The great building epoch (EN II-MN I) has been referred to by researchers as an economic and ritual expansion phase in which elaborately decorated pottery was deposited in front of the megalithic tombs. Within the same period of expansion we see the building of causewayed enclosures/camps, cult houses and an increased ritual focus on wetland hoarding (Andersen 2000, 16-38).

After the ritual expansion phase, the construction of monumental tombs stopped but their use however continued. This ongoing practice of megalithic entombment continued on the Danish Islands well into the Bronze Age. Significant is the combined period of MN III/IV (the Bundsø/Lindø phase), which holds about 55 % of the Middle Neolithic megalithic pottery from the Danish Islands (Ebbesen 1975, 123, fig. 103). In Jutland a marked reduction in the amount of pottery deposited in megalithic tombs took place after MN I (Ebbesen 1978, 100-103, figs. 94-95).

The final Funnel Beaker phase (MN V) is defined by the St. Valby pottery style that traditionally has been dated between 2900-2800 BC whereupon the Single Grave horizon begins in central and western Jutland (cf. fig. 1). In contrast, the St. Valby phase ran from c 2950-2650 BC in easternmost Jutland and on the Danish Islands as indicated by the available 14C dates (cf. Midgley 1992, 218-219, fig. 71). A similar long duration of the late Funnel Beaker phase apply to Bornholm and Scania where we see the existence of local developments in the form of the Vasagård and Grødby styles and the Stävie/Karlsfält group respectively (Larsson

Years BC (calibrated)	Name of culture and period					Grave type, building period	Type of enclosure, building period
--- 2450 --- --- 2600 ---	Single-Grave/ Battle-Axe Culture	Middle Neolithic B (MN B)	III	Upper Graves	Per. 5	Barrows, stone cists and flat graves	
			II	Ground Graves	Per. 4-5		
	Pitted-Ware		I	Bottom Graves	Per. 1-3		Palisade enclosures
--- 2800 --- --- 2900 ---	-------	Middle Neolithic A (MN A)	V	St. Valby			
--- 3000 ---	Funnel-Beaker Culture		III	Bundsø			
--- 3100 ---			II	Blandebjerg			
--- 3200 ---			Ib	Klintebakke		Passage graves	Causewayed camps
--- 3300 ---			Ia	Troldebjerg		Enlarged dolmens	
		Early Neolithic (EN)	C	Fuchsberg / Virum		Early dolmens	
--- 3500 ---			B	Volling		Earthen long barrows	
			A	Oxie			

Figure 1. Traditional chronological scheme covering the Early and Middle Neolithic in Denmark (from P.O. Nielsen 2004: table 1).

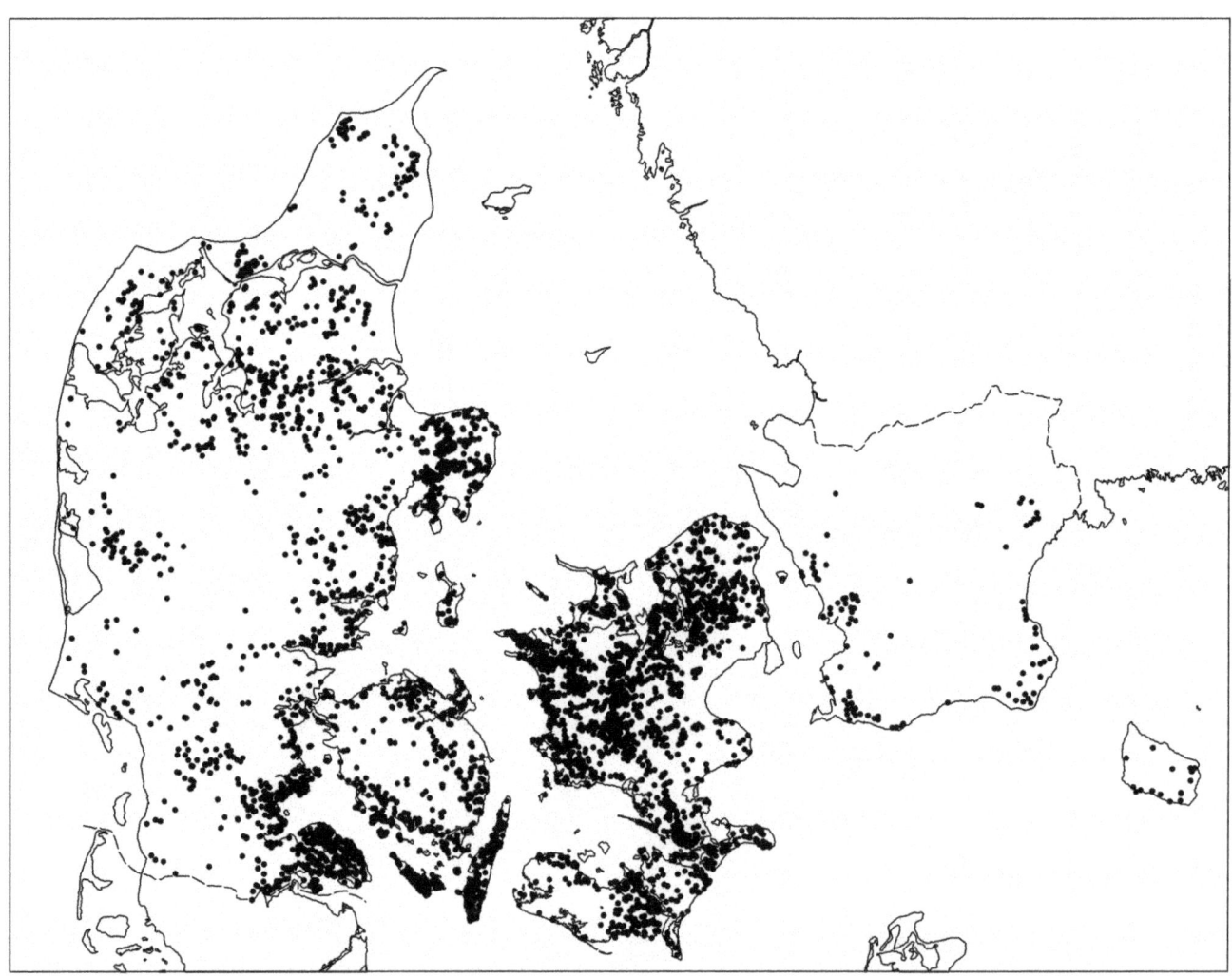

Figure 2. Megalithic tombs in Denmark and Scania. Source: The national database 'Sites and Monuments', the Heritage Agency of Denmark and Tilley 1996: fig. 3.18. Geographic map: the research programme 'Settlement and Landscape'.

1982, 100; Edenmo *et al.* 1997, 141-9; P.O. Nielsen 1997; Kaul *et al.* 2002, 131, 136; M. Larsson 2006, 73-78). St. Valby pottery makes up less than 10 % of the total amount of Middle Neolithic megalithic pottery from the Danish Islands. In the megalithic tombs of Jutland the frequency is c 7 % in southern Jutland and less than 5 % in northern Jutland (Ebbesen 1975, figs. 102, 103; 1978, 102-3, fig. 94). Instead, flint axes played an increasing role during the later Funnel Beaker period and succeeded pottery as the prime deposition object in megalithic tombs as well as in wetland hoards (Andersen 2000, 38).

With the emergence of the Single Grave culture c 2850 BC on the Jutland Peninsula (fig. 3), a uniform burial practice was introduced with individual interments covered by a low mound and battle-axes as the most significant grave good. Close to 2400 single graves are known from Jutland (Hübner 2005, 60). Outside the Single Grave core area megalithic entombments continued, in particular on the Danish Islands, in Mecklenburg-Vorpommern and in eastern Schleswig-Holstein (Ebbesen 2006, 123, fig. 93).

On the Danish Islands (incl. Bornholm) and in Scania, the Funnel Beaker tradition continued in the form of a prolonged MN V phase and a permanent use of megalithic tombs. Single Grave culture artefacts were gradually incorporated in the late Middle Neolithic material culture as the tanged arrowheads related to the Pitted Ware complex. A considerable amount of these arrowheads have been found in East Danish megalithic tombs, whereas no individual Pitted Ware graves have been recorded in Denmark (Iversen 2010, 15-8, 26). In a South Scandinavian context, the Pitted Ware complex is found in the north-eastern parts of Denmark and scattered along the South and West Swedish coast.

Farming, settlement size and the use of alternative subsistence strategies

In general, the Funnel Beaker economy was based on agriculture and husbandry, primarily cattle (Rowley-Conwy 1984, 89-92; Midgley 1992, 372). Wild species only played a minor role though coastal sites contain larger amounts of marine resources. Still, wild animal bones only make up about six percent of the fauna bone material on these sites (Nyegaard 1985, 454). Furthermore analyses of the 15N isotope values from human skeletal remains from the Danish Neolithic, suggest a diet consisting of primarily vegetables and meat, and to a smaller degree freshwater fish (Richards and Koch 2001; Bennike and Alexandersen 2007, 132). Similarly, a decrease in 13C isotope values indicates that about 70 % of the marine dominated hunter-gatherer diet was replaced by farming produce at the Mesolithic/ Neolithic transition (Noe-Nygaard 1988, 94). Based upon fauna material composition and geographical location, some Funnel Beaker sites must be conceived as more or less specialised hunting sites though. All such sites date from the Early Neolithic and beginning of the Middle Neolithic (Skaarup 1973, 118-133).

At the beginning of the Early Neolithic, the settlement pattern was dispersed with smaller sites about 500-700 m2 located on light soils. Houses are small, 10-15 m long two-aisled structures, and represent mobile and short-term occupation based on slash and burn agriculture. From the onset of the Middle Neolithic (c 3300 BC) the settlement pattern changes towards larger units of longer duration with sites exceeding 30,000 m2. Focus was on agricultural intensification including grain cultivation and the establishment of permanent fields and commons. This development is extended during the final Funnel Beaker phase (MN V) in which settlements grew bigger and in some cases covered more than 70,000 m2. Also the houses got bigger but kept a two-aisled layout now measuring c 18-22 m in length (Madsen 1982, 222-231; P.O. Nielsen 1993, 92-95; 1999, 150-155; Rasmussen *et al.* 2002, 5; Johansen 2006, 208-14).

However, one must take some precautions concerning the size of the larger settlements since these are defined on the basis of stray finds. Widespread settlement finds might as well represent various smaller sites as one large. Considering the prolonged duration of the final Funnel Beaker phase (MN V) it is reasonable to expect more sites and higher concentration of finds compared to the earlier Middle Neolithic phases. That said, the archaeological record still indicates a centralisation of the settlement pattern but unfortunately house remains are generally sparse due to the lack of substantial excavations on many of the larger sites.

The development in the Funnel Beaker settlements can best be describes as one of centralisation and enlargement from the late Early Neolithic through the early Middle Neolithic. The causewayed enclosures probably functioned as a trigger for the development of large nucleated settlements since Middle Neolithic settlements are often located on former causewayed enclosures. The reuse of the causewayed enclosure sites included the re-digging of ditches, a practice that can even be observed on some larger Pitted Ware sites (Larsson 1982, 84; Andersen 1997, 118-124; Nielsen 2004, 24-8, 30). Connected to the centralisation of the late Funnel Beaker settlements is the emergence of palisaded enclosures located on Zealand, Scania and Bornholm. Flint axes and the debris from the axe production were deposited at several of the palisaded enclosures together with burnt flint tools, features that have also been observed at other large Middle Neolithic sites. The probable ritual character of the palisaded enclosures is underlined by the fact that several were destroyed by fire (Svensson 2002, 46-50; Nielsen 2004, 26-28; Brink 2009, 31-42, 343-4, fig. 1).

Single Grave and Pitted Ware alternatives

House remains related to the Single Grave culture (2850-2350 BC) are known from the Jutland Peninsula and have mainly been identified as shallow depressions filled with settlement refuse or light post built structures. Settlement sites are additionally small and dispersed consisting

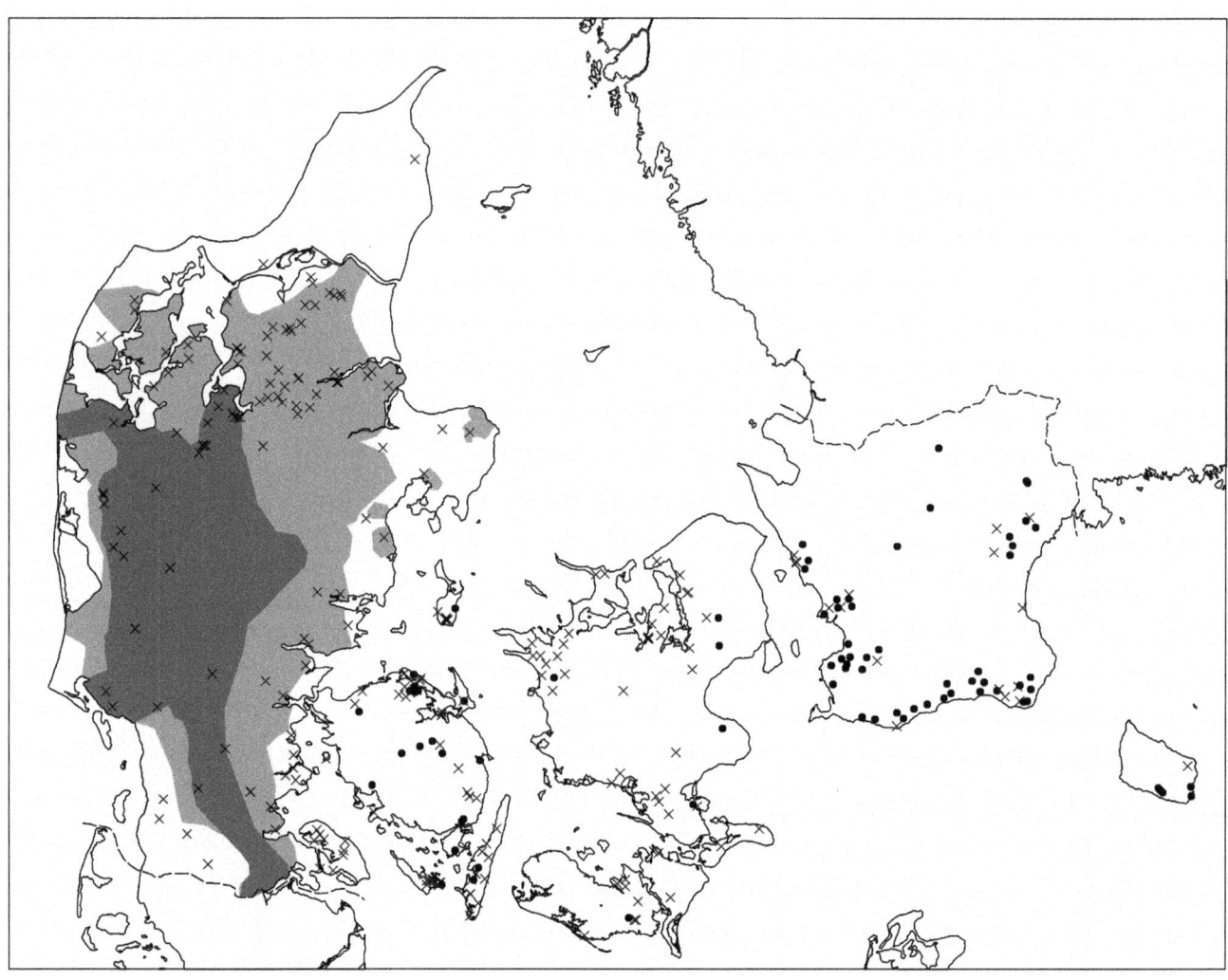

Figure 3. Single Grave and Battle Axes culture graves in Denmark and Scania (dot). Grey colouring: distribution of Jutland Single Graves. Dark grey: initial phase c 2850-2800 BC. Cross: megalithic tombs with Single Grave/Battle Axe culture finds. Source: F.O. Nielsen 1987: fig. 1; L. Larsson 1993: fig. 57; Hübner 2005: figs. 419-20, 470-6; Ebbesen 2006: find lists 1-3, 4-5; AUD and 'Sites and Monuments', the Heritage Agency of Denmark.

of individual households located on sandy soils. The subsistence economy is badly elucidated but includes husbandry with a supposed focus on cattle, sheep and, on a smaller scale, the cultivation of cereals; mainly barley (Odgaard 1986; Simonsen 1987; Odgaard and Rostholm 1988; Liversage 1988, 120-123; Andersen 1998, 126-127; Nielsen 1999, 156; Klassen 2005, 34-39; Ebbesen 2006, 224). As an exception, continuity in settlement pattern and house constructions from the late Funnel Beaker to the early Battle Axe culture phase (c 2800-2500 BC) can be followed on Bornholm. The houses are rectangular two-aisled structures measuring around 20 m in length continuing the late Funnel Beaker tradition (Nielsen 1999, 154-156). Also from the Malmö region, regular houses dating between c 2800-2350 BC have been found (Brink 2009, 268-275).

Even though farming played a role in the Single Grave economy, the focus on sandy soils and short-lived dispersed settlements made up of individual light built houses, show a significant change in the lifeways of the Neolithic. A mobile

lifestyle combined with herding played an increasing role while the use of megalithic tombs was scaled down in favour of single interments.

Another break away from the otherwise intensified Neolithic lifestyle represented by the late Funnel Beaker culture is the occurrence of Pitted Ware sites in South Scandinavia (c 3000-2450 BC). Traditionally, the Pitted Ware complex has been defined from the pottery found on some larger sites on the East Swedish mainland. In South Scandinavia, however, tanged arrowheads and cylindrical blade cores (used for the production of arrowheads) constitute the most frequently observed Pitted Ware elements (Lidén 1940, 90; Becker 1951, 180-192). Pitted Ware sites mostly reflect a mobile lifestyle with seasonal or short-lived camps closely connected to the coastal zones, fjords and watercourses. Sites are almost exclusively distributed in the northern and north-eastern parts of Denmark. Only north-eastern Jutland has larger sites with traces of all-year occupation (Rasmussen 1991; Iversen 2010, 11-5, fig. 5).

Houses are rare even if we consider the whole of Scandinavia and no secure Pitted Ware houses or huts have been recorded in Denmark. In spite of the high concentration of Pitted Ware sites in eastern-middle Sweden (more than 200) the majority are without house constructions. In the case of houses, we only see minor huts measuring about three-square meters (Malmer 2002, 97-126).

The location and organization of Pitted Ware sites differ markedly from the late Funnel Beaker settlement pattern consisting of large sites with solid built longhouses as seen on Bornholm.

Differences in site organization are also reflected in the subsistence economy though the archaeological evidence is sparse. When discussing the Pitted Ware economy, the East Jutland site of Kainsbakke plays an important role due to its content of preserved animal bones and shells (Rasmussen 1984; 1991). The fauna bone material shows year-round occupation and thereby gives a unique glimpse of the Pitted Ware economy without the often-needed precautions regarding specialised or seasonal determined strategies. The site reveals a broad-spectrum economy consisting of husbandry, hunting, fishing and gathering with a certain importance attached to the marine resources (including shellfish), which have been extensively exploited. Concerning the use of agriculture no direct traces have been found on Kainsbakke even though farming is indicated by the existence of grindstones (Rasmussen 1991, 45).

There are indications of agriculture from the South Swedish Jonstorp sites where grindstones and pottery with grain impressions have been found (Lidén 1940, 146, 189-193). Other sites from Denmark show that shellfish played a substantial part of the diet and was supplemented by game, fish and husbandry (Marseen 1953,113-116). On the small island of Hesselø in the Kattegat sea, the food strategy was specialised on the exploitation of seal (Becker 1951, 165; Møhl 1971, 308), a situation that mirrors the general picture known from Gotland (Eriksson 2004).

In spite of the fact that only few sites contain preserved faunal material and no sites have produced dwelling structures, it is clear that real differences existed in the way Funnel Beaker and Pitted Ware people structured and dealt with daily life. The gathering of molluscs and the exploitation of marine resources in general played a substantial role in the Pitted Ware diet; aspects of the economy which are clearly mirrored in the coastal location of the settlements. Supporting a marine diet, hunting, husbandry and to a smaller extent agriculture were all elements of a mixed South Scandinavian Pitted Ware subsistence economy.

The cease of the Funnel Beaker culture and the 'de-Neolithisation' of South Scandinavia

No doubt far-reaching changes took place in the first half of the third millennium BC. After the consolidation and

intensification of agricultural life including a marked increase in settlement size, the building of solid timber houses and the construction of enclosures, this development stopped. The palisaded enclosures of MN V can be seen as a final large-scale manifestation within the context of the Funnel Beaker tradition.

Even though it may be tempting to conceive the described changes as rapid and revolutionary transformations of society caused by conflicting cultural groups, the change of lifestyle is best explained as the result of new impulses transmitted through contact networks combined with local developments. The interaction of foreign influx and local conditions explains the observed cultural change and the emergence of both the Single Grave culture (Damm 1993, 201-203; Klassen 2005, 45-47) and the Pitted Ware complex (Iversen 2010, 23-27). Behind these contacts, which encouraged new ways of Neolithic life and changes in the material culture, we must expect personal relations and the movement of individuals. That highly mobile groups or individuals were travelling within Neolithic Europe has been confirmed by isotope and DNA analysis from various places (Price *et al.* 2004; Evans *et al.* 2006; Malmström *et al.* 2009).

However, cultural and subsistence economic changes happened at various tempi and were more or less profound depending on local preconditions. In parts of Jutland we see a relatively rapid transition from the late Funnel Beaker culture (MN V) to a new wide-ranging Corded Ware tradition (c 2850 BC). This cultural shift involved the incorporation of former sparsely populated areas covering the sandy soils of South-West Jutland, dispersed settlements, herding, an increased focus on battle-axes and the use of curved cord ornamented beakers. Additionally, individual interments underneath low mounds were part of the new Corded Ware/Single Grave organization at the expense of the communal megalithic tombs.

In the eastern parts of Denmark (incl. easternmost Jutland) and Scania, the late Funnel Beaker culture continued until c 2650 BC with some regional developments in pottery style. Megalithic tombs were the predominant burial form throughout the Neolithic. Already from c 3000 BC Pitted Ware elements were incorporated in the gradually disintegrated Funnel Beaker tradition. The reuse of megalithic tombs is in particular visible on the Danish Islands where Pitted Ware tanged arrowheads and certain variants of Single Grave battle-axes and beakers were incorporated in the existing burial custom illustrating the gradual disintegration of the Funnel Beaker culture. In Scania and on Bornholm, the Swedish-Norwegian Battle Axe culture constituted a particular variant of the widespread Corded Ware complex (cf. fig. 3).

Different contact networks divided South Scandinavia in the first centuries of the third millennium BC. In spite of marked local differences the result was the same: a downgrading of farming in favour of mobile resources like herding and hunting strategies combined with a disintegration of the settlement structure. With some precautions I will describe this process as one of 'de-Neolithisation'. By using this term I wish to highlight the point that the adoption of agricultural economy and Neolithic lifestyle was not a straightforward process that ended with the cultivation of the first crops around 4000 BC. This is most clearly illustrated by the revival of hunter-fisher-gatherer strategies represented by the Pitted Ware complex, which in a South Scandinavian setting included part-time agriculture and use of megalithic tombs. As contact networks and people introduced new ideas into the existing social and economic structures of the Funnel Beaker culture, Neolithic life changed and agriculture was downgraded in favour of more mobile subsistence strategies. First with the beginning of the Late Neolithic (c 2350 BC) did this picture change and even longer timber houses were built than previously seen mirroring 'chieftain halls' and the emergence of a distinct social elite.

References

Andersen, N. H. 1997. *The Sarup Enclosures. The Funnel Beaker Culture of the Sarup site including two causewayed camps compared to the contemporary settlements in the area and other European enclosures.* Jutland Archaeological Society Publications XXXIII:1. Århus, Jutland Archaeological Society.

-2000. Kult og ritualer i den ældre bondestenalder. *Kuml*, 13-57.

Andersen, S. Th. 1998. Pollen analytical investigations of barrows from the Funnel Beaker and Single Grave Cultures in the Vroue area, West Jutland, Denmark. *Journal of Danish Archaeology* 12,107-132.

AUD – Arkæologiske Udgravninger i Danmark 1984-2005. København: Rigsantikvarens Arkæologiske Sekretariat/ Kulturarvsstyrelsen.

Becker, C. J. 1951. Den grubekeramiske kultur i Danmark. *Aarbøger for nordisk Oldkyndighed og Historie* 1950, pp. 153-274.

Bennike, P. and V. Alexandersen 2007. Population Plasticity in Southern Scandinavia. From Oysters and Fish to Gruel and Meat. In M. N. Cohen and G. M. M. Crane-Kramer (eds.), *Ancient Health. Skeletal Indicators of Agricultural and Economic Intensification,* 30-148. Gainesville, University Press of Florida.

Brink, K. 2009. *I palissadernas tid. Om stolphål och skärvor och sociala relationer under yngre mellanneolitikum. Malmöfynd nr. 21.* Malmø, Malmö Museer.

Damm, C. 1993. The Danish Single Grave Culture – Ethnic Migration or Social Construction? *Journal of Danish Archaeology* 10, 199-204.

Dehn, T. and S. I. Hansen 2006. Birch bark in Danish passage graves. *Journal of Danish Archaeology* 14, 23-44.

Dehn, T., Hansen, S. I. and F. Kaul 2000. Klekkendehøj og Jordehøj. Restaurering og undersøgelser 1985-90. Stenaldergrave i Danmark 2. København, Nationalmuseet and Skov- og Naturstyrelsen.

Ebbesen, K. 1975. *Die jüngere Trichterbecherkultur auf den dänischen Inseln. Arkæologiske Studier II.* København, Akademisk Forlag.

-1978. Tragtbægerkultur i Nordjylland. Studier over jættestuetiden. In *Nordiske Fortidsminder, Serie B, 5.* København, Det Kongelige Nordiske, Oldskriftselskab.

-2006. *The Battle Axe Period/Stridsøksetid.* København, Attika

Edenmo, R., Larsson, M., Nordqvist, B. and Olsson, E. 1997. Gropkeramikerna – fanns de? Materiell kultur och ideologisk förändring. In M. Larsson and E. Olsson (eds.), *Regionalt och interregionalt. Stenåldersundersökningar i Syd- och Mellansverige. Arkeologiska undersökningar. Skrifter 23,* 135-213. Stockholm, Riksantikvarieämbetet.

Eriksson, G. 2004. Part-time farmers or hard-core sealers? Västerbjers studied by means of stable isotope analysis. *Journal of Anthropological Archaeology* 23, 135-162.

Evans, J.A., Chenery, C.A. and Fitzpatrick, A.P. 2006. Bronze Age childhood migration of individuals near Stonehenge, revealed by strontium and oxygen isotope tooth enamel analysis. *Archaeometry* 48 (2), 309-321.

Hübner, E. 2005. Jungneolithische Gräber auf der Jütischen Halbinsel. Typologische und chronologische Studien zur Einzelgrabkultur. In *Nordiske Fortidsminder, Serie B, 24.* København, Det Kongelige Nordiske, Oldskriftselskab.

Iversen, R. 2010. In a worlds of worlds. The Pitted Ware complex in a large scale perspective. *Acta Archaeologica* 81, 5-43.

Jensen, J. 2001. *Danmarks Oldtid. Stenalder 13.000 – 2.000 f. Kr.* København, Gyldendal.

Johansen, K. L. 2006. Settlement and Land Use at the Mesolithic-Neolithic Transition in Southern Scandinavia. *Journal of Danish Archaeology* 14, 201-223.

Kaul, F., Nielsen, F. O. and Nielsen, P. O. 2002. Vasagård og Rispebjerg. To indhegnede bopladser fra yngre stenalder på Bornholm. *Nationalmuseets Arbejdsmark* 119-138.

Klassen, L. 2005. Refshøjgård. Et bemærkelsesværdigt gravfund fra enkeltgravskulturen. *Kuml*, 17-59.

Larsson, L. 1982. A Causewayed Enclosure and a Site with Valby Pottery at Stävie, Western Scania. *In Meddelanden från Lunds Universitets Historiska Museum 1981-82,* 65-107.

-1992. Settlement and Environment during the Middle Neolithic and Late Neolithic. In: L. Larsson, J. Callmer and B. Stjernqvist (eds.), *The Archaeology of the Cultural Landscape. Fieldwork and Research in a South Swedish Rural Region. Acta Archaeologica Lundensia,* 91-159. Lund, Institute of Archaeology and the Historical Museum.

Larsson, M. 2006. *A Tale of a Strange People. The Pitted Ware Culture in Southern Sweden. Kalmar Studies in Archaeology 2.* Lund, University of Lund.

Lidén, O. 1940. *Sydsvensk Stenålder II. Strandboplatserna i Jonstorp.* Lund, Gleerupska.

Liversage, D. 1988. Mortens Sande 2 – A Single Grave Camp Site in Northwest Jutland. *Journal of Danish Archaeology* 6, 101-124.

-1992. *Barkær. Long Barrows and Settlements. Arkæologiske Studier Vol. IX.* København, Akademisk Forlag.

Madsen, T. 1982. Settlement Systems of Early Agricultural Societies in East Jutland, Denmark, A Regional Study of Change. *Journal of Anthropological Archaeology* 1(3),197-236.

Malmer, M. P. 2002. The Neolithic of South Sweden. TRB, GRK, and STR. Stockholm, The Royal Swedish Academy of Letters History and Antiquities. Stockhom, The Royal Swedish Academy.

Malmström, H., Gilbert, M. T. P., Thomas, M. G., Brandström, M., Storå, J., Molnar, P., Andersen, P. K., Bendixen, C., Holmlund, G., Götherström, A. and Willerslev, E. 2009. Ancient DNA Reveals Lack of Continuity between Neolithic Hunter-Gatherers and Contemporary Scandinavians. *Current Biology* 19(20), 1758-1762.

Marseen, O. 1953. Fangstfolk på Selbjerg. *Kuml,* 102-120.

Midgley, M. 1992. TRB Culture. *The First Farmers of the North European Plain*. Edinburgh, Edinburgh University Press.

Møhl, U. 1971. Fangstdyrene ved de danske strande. *Kuml,* 97-329.

Nielsen, F. O. 1989. Nye fund fra stridsøksetiden på Bornholm. In Larsson, L. (ed.), *Stridsyxekultur i Sydskandinavien. University of Lund Institute of Archaeology Report Series 36,* 89-101. Lund, University of Lund.

Nielsen, P. O. 1993. Bosættelsen. In S. Hvass and B. Storgaard (eds.), *Da klinger i muld... 25 års arkæologi i Danmark (Digging into the Past, 25 years of archaeology in Denmark),* 92-95. København and Århus, Det Kongelige Nordiske Oldskriftselskab and Jysk Arkæologisk Selskab.

-1997. Keeping Battle-Axe People away from the Door. Neolithic House-Sites at Limensgård and Grødbygård, Bornholm. In D. Król, (ed.), *The Built Environment of Coast Areas during the Stone Age. The Baltic Sea-Coast Landscapes Seminar 1,* 196-208. Gdańsk, Regional Centre for Studies and Preservation of Built Environment.

-1999. Limensgård and Grødbygård. Settlements with house remains from the Early, Middle and Late Neolithic on Bornholm. In C. Fabech and J. Ringtved (eds.), *Settlement and Landscape,* 149-165. Højbjerg, Jutland Archaeological Society.

-2004. Causewayed camps, palisade enclosures and central settlements of the Middle Neolithic in Denmark. *Journal of Nordic Archaeological Science* 14, 19-33.

Noe-Nygaard, N. 1988. δ13C-values of dog bones reveal the nature of changes in man's food resources at the Mesolithic-Neolithic transition, Denmark. *Chemical Geology, Isotope Geoscience,* 73, 87-95.

Nyegaard, G. 1985. Faunalevn fra yngre stenalder på øerne syd for Fyn. In J. Skaarup (ed.), *Yngre stenalder på øerne syd for Fyn. Meddelelser fra Langelands Museum,* 426-457. Rudkøbing, Langelands Museum.

Odgaard, B. V. 1986. Enkeltgravskulturens miljø i Vestjylland belyst gennem pollendiagrammer. In C. Adamsen and K. Ebbesen (eds.), *Stridsøksetid I Sydskandinavien. Arkæologiske Skrifter 1,* 194-195. København, Forhistorisk Arkæologisk Institut Københavns Universitet.

Odgaard, B. V. and Rostholm, H. 1988. A Single Grave Barrow at Harreskov, Jutland. Excavation and Pollen analysis of a Fossil Soil. *Journal of Danish Archaeology* 6, 87-100.

Price, T. D., Knipper, C., Grupe, G. and Smrcka, V. 2004. Strontium Isotopes and Prehistoric Human Migration, the Bell Beaker Period in Central Europe. *European Journal of Archaeology* 7(1) 9-40.

Rasmussen, L. W. 1984. Kainsbakke A47, A Settlement Structure from the Pitted Ware Culture. *Journal of Danish Archaeology* 3, 83-98.

-1991. Kainsbakke. En kystboplads fra yngre stenalder. In L.W. Rasmussen and J. Richter, *Kainsbakke,* 9-69. Grenaa, Djurslands Museum, Dansk Fiskerimuseum.

Rasmussen, P., Bradshaw, E. and Andersen, N. H. 2002. Danmarks tidlige landbrug. – Et nyt forskningsprojekt og et overraskende fund af laminerede søsedimenter. Geologi. Nyt fra GEUS 2002, (2) 2-6.

Richards, M. P. and Koch, E. 2001. Neolitisk kost. Analyser af kvælstof-isotopen 15N i menneskeskeletter fra yngre stenalder. *Aarbøger for nordisk Oldkyndighed og Historie,* 7-17.

Rowley-Conwy, P. 1984. Mellemneolitisk økonomi i Danmark og Sydengland. Knoglefundene fra Fannerup. *Kuml,* 77-111.

Simonsen, J. 1987.Settlements from the Single Grave Culture in NW-Jutland. A Preliminary Survey. *Journal of Danish Archaeology* 5, 135-151.

Skaarup, J. 1973. *Hesselø – Sølager. Jagdstationen der südskandinavischen Trichterbecherkultur. Arkæologiske Studier I.* København, Akademisk Forlag.

Svensson, M. 2002. Palisade Enclosures – The Second Generation of Enclosed Sites in the Neolithic of Northern Europe. In A. Gibson (ed.), *Behind Wooden Walls, Neolithic Palisaded Enclosures in Europe,* 28-58. Oxford, BAR International Series 1013.

Thomas, J. 1999. *Understanding the Neolithic. A revised second edition of Rethinking the Neolithic.* London, Routledge.

Tilley, C. 1996. *An Ethnography of the Neolithic. Early Prehistoric Societies in Southern Scandinavia.* Cambridge, Cambridge University Press.

CHAPTER 4

'IN-BETWEEN':
RE-THINKING THE CONTEXT OF THE BRITISH MESOLITHIC-NEOLITHIC TRANSITION
PRELIMINARY THOUGHTS

Irene Garcia Rovira

irene.g.rovira@gmail.com,
University of Manchester,
Department of Archaeology.

Abstract: *Archaeologists often claim to have found alternative ways of thinking about the past. Yet, old ideas 'persist and recur' (Trigger 1989, 8), and primary projections often play a central role in the way in which we conceptualise certain archaeological contexts (see Tilley 1998). One such example can be clearly defined in the literature of the Mesolithic-Neolithic transition. Here, traditional models (e.g. Pumpelly 1908; Vavilov 1926; Childe 1925) have, to a greater extent, to structure our understanding of this period of change. The transition continues to be framed within origin/dispersal models.*

Whilst this might be the case, researchers strive to find ever more suitable methodological and theoretical strategies that allow obtaining the complexities of this context of change. In recent years, many archaeologists have opted for the development of studies focused on reduced scales of analysis, seeking to remove the generalisations inherent in traditional models. Moreover, a great amount of effort has been placed at implementing notions of agency to obtain 'particularistic histories of change' (Whittle 2007, 626) which go beyond accounts that underestimate the role of Mesolithic populations .

This article offers a primary reflection on some of the consequences that have arisen with the application of the theoretical strategies mentioned above. It is argued that, whilst agency and practice theory challenge traditional accounts of this context of change, by the very nature of the theoretical frameworks, these approaches fail to characterise the transformative nature of interstitial spaces. Hybridity discourses might offer new interpretive avenues to a context, which is essentially characterised by a spread triggered in cultural encounters of one kind or another.

This discussion is taken to an end by opening up the following question: if we are ready to accept that this context should be considered as multiple transitions (e.g. Whittle 2007; Whittle et al. 2011), should we continue to link these events to a single process? Rather than offering an exhaustive discussion on the propositions stated above, this article seeks to open new interpretive avenues, which can be pursued in future research.

Keywords: *Mesolithic-Neolithic transition, agency theory, practice theory, cultural encounters, post-colonial theory, hybridity, and entanglement*

Introduction

> *'Our understanding of the [transition] requires coherent and subtle theorisation. The evidence will not speak for itself'. (Thomas 2008, 59).*

The British Mesolithic-Neolithic transition has been subject to no less than fifty years of continuous debate (e.g. Piggott 1954; Case 1969); to enduring discussions both sustained and questioned by research directions mutually endeavouring to define the character of this process (or processes). However, whilst the debate has been constantly evolving over time, this has been punctuated by different waves of research that have triggered fresh interest in the topic. In recent years, the transition has been revisited in light of a new battery of data.

Over the last decade, palaeodiet specialists using isotopic analysis have alerted us to the absence of marine protein found in human skeletal examples dating to c. 4000 cal BC (e.g. Schulting and Richards 2002). Whilst not without criticism (see Milner et al. 2003), this evidence has been central to discussing changing subsistence patterns and food taboos in the context of the British Mesolithic-Neolithic transition (see Schulting and Hedges 2003; Thomas 2003). Other authorities have reminded us about the complexity of such processes, pointing to the level of variability that one can encounter when examining the evidence regionally (e.g. Cummings 2007). The significance of comparative dating projects, such as those developed with the application of Bayesian techniques (e.g. Whittle et al. 2011) is beginning to crystallise, reconfiguring the narratives hitherto produced for this archaeological context. Alongside this, new research directions have developed, resulting from the influence of particular approaches arising within post-processual discourse (e.g. Tilley 2007).

Although increasing interest in this topic has encouraged the development of an overwhelming array of literature and an unprecedented corpus of data, discrepancies in the mechanisms which gave rise to the period of change have remained pervasive and so, as Whittle (2003) has pointed out, the transition has remained 'stubbornly and frustratingly unclear' (Whittle 2003, 150). Recent examinations of this socio-historical context have been largely produced to either substantiate the validity of polarised models (indigenous/colonisation model) (e.g. Sheridan 2003, 2004; Thomas 1988, 2003, 2008), or to suggest the possibility of a more complex situation, which can only be assessed regionally (e.g. Whittle et al. 2011; Cummings and Harris 2011).

Whilst denouncing the inherent dangers of constructing 'black and white' (Garrow and Sturt 2011) views of the transition, the latter group has been particularly inclined to consider theoretical strategies needed to challenge existing narratives. The examination of human agency and the definition of studies focused on small-scale analysis have been prominent choices selected to obtain the 'grey' of this archaeological context (see Warren 2007; Whittle et al. 2011; Cummings and Harris 2011). However, can these theoretical and methodological reorientations shed light into the context within which the so-called transition took place?

To discuss this question, this article begins by exploring the ways in which agency theory (e.g. Dobres and Robb 2000) has been used to obtain 'particularistic histories of change' (Whittle 2007, 626) recently sought in the context of the British Mesolithic-Neolithic transition. Whilst insights on practice theory (Bourdieu 1977, 1990; Giddens 1979, 1984), have helped to provide better understandings on the mechanisms by which, for instance, alien cultural practices are internalised and transformed within specific socio-cultural milieus, they fail to characterise the phenomena triggered in interstitial contexts. If, as Thomas (2008) (see quote above), suggests better understandings of the transition are to be attained through thorough theorisation, would it not perhaps be necessary to obtain better insights into the kinds of phenomena taking place in the spaces 'in-between' (Bhabha 1994)? Following on from this concern, in this section, the article will seek to ground theoretical resources that can be used to trigger new understandings about the transition.

The recent emphasis that has been placed on understanding this archaeological context as multiple transitions (e.g. Whittle 2007; Whittle et al. 2011), brings about the need to consider the following question: is it still possible to argue for the existence of a common underlying theme which links these events into a single process? Whilst this question has been recently addressed by Hodder (2012) in the context of the introduction of farming in the Middle East, here I would like to draw attention to changing social relationships triggered by what, using Marxist terminology, would be defined as changing modes of production. Is it possible that the directionality observed in the processes of diffusion, implicitly defined in the context of the British Mesolithic-Neolithic transition, would have been triggered by changing social relationships within and amongst different social groups? This question leads to the need to develop an approach, which reconciles the events of transformation and of cultural production defined at the micro-scale with new social understandings induced in the context of changing social relationships of production.

Far from developing an exhaustive account based on the discussion followed in this article, I will instead seek to open new interpretive avenues which can be pursued in future research.

From 'within' to the 'in-between': re-thinking the context of the transition

'In a similar vein, as archaeologists for the last twenty years have struggled to develop alternative discourses and epistemologies to dislocate their practices from constraining empiricist philosophies, we are now faced with the challenge of discussing new socio-political agendas that go beyond narrow histories of identity. To start thinking that against the 'absolutism of the pure' (Rushdie 1982), the well-proven solution of returning to an ethnic and/or national primordialism, 'there is another different option: the theorisation of creolisation, [...] mestisaje and hybridity (Gilroy 1993, 2)'. (Olsen 2001, 53-54).

Recent reviews, which have focused on theoretical as well as methodological strategies traditionally used to examine the transition, have generally concurred in arguing the necessity of exploring agency and establishing regional studies in order to obtain a fuller picture of the contingencies of this context of change. The latter initiative explicitly draws upon the risks of drawing a single model that throws the multiplicity of responses potentially generated by the introduction of new subsistence regimes and material forms during the transition into a pool of homogenisation. However, the underlying reasons triggering the application of agency theory are twofold and demand critical consideration.

At first sight, it could be suggested that calls for agency have been in reaction to the limitations set by traditional models (e.g. Childe 1925; Ammerman and Cavalli-Sforza 1984), which did not recognise the role that Mesolithic populations played in the aforementioned process of change. In this context, attention has been drawn to the danger of defining simplistic scenarios in which Neolithic populations imposed new values, customs and so forth on Mesolithic groups. In order to avoid the depiction of hunter-gatherers simply acting as cultural dupes (Borić 2005, 85), these proposals view agency as a means of considering situations in which hunter-gatherers were actively choosing whether 'to farm or to not farm', or 'to build or not to build' (Fewster 2001; Warren 2007, 318) and so on. Here then, agency might be characterised as homologous to 'intentional action; rational action; conscious practice' (Dornan 2002, 304); it is, on the whole, discursive.

Even though these approaches adeptly consider the role of hunter-gatherer populations as agents in this process of change, they have developed, adopting a view of agency, which has often been questioned in the wider arena of archaeological theory (e.g. Barrett 2000; Dornan 2002). In this context, attention has been drawn to the need to consider agency as 'a socially significant quality of action rather than being synonymous with, or being reducible to, action itself' (Dobres and Robb 2000, 8). But not all accounts of agency in the context of the transition can be said to depart from the same predicates.

With the benefit of hindsight, it might be argued that the introduction of the issue of agency within these debates also corresponds to the infiltration of wider theoretical discussions whose origins are to be found, predominantly in the application of 'practice theory' (e.g. Bourdieu 1977, 1990; Giddens 1979, 1984) in archaeology. In this case, the politics of representation suggested go beyond situations in which specific social groups actively adopted or rejected cultural phenomena encountered in situations of intercultural interaction. Instead, these theoretical insights endeavour to take into account the ways in which novel cultural practices are internalised and transformed within specific socio-cultural settings. A good example of this position – although perhaps one, which does not draw directly from the introduction of 'agency theory' - can be found in the ways in which Thomas (e.g. 1988, 2003, 2008) has defined the main factors that gave rise to what we describe as the transition. Although in some instances (e.g. Thomas 2006a, 56), he describes the changes occurring at the end of the 5th millennium BC as moments of hybridisation and mixing, his approach largely posits a context in which hunter-gatherers 'adopted' Neolithic traits whilst maintaining aspects of their Mesolithic lifestyles (Cummings and Harris 2011, 361-2). On this occasion, agency is not seen as simplistic intentionality, but as a creative force engendered in social practice. However, it is worth enquiring whether or not this approach is sufficient to offer a fuller picture of the complex nature of the changes that took place during the transition.

Despite the difficulties encountered in establishing an agreement on the nature and character of the transition, it has been commonly agreed that this archaeological context echoes processes triggered in interstitial contexts. In the case of the British Mesolithic-Neolithic transition, it has been generally acknowledged that the emergence of 'Neolithic things and practices' (Whittle et al. 2011) in different regions of the British Isles corresponds to phenomena occurring through the interaction of different social groups. Scenarios of this kind were, in traditional paradigms (e.g. Childe 1925), described as resulting from processes of diffusion; processes openly discussed at a more theoretical level (e.g. Boas 1896; Montelius 1905), and central to explanations of cultural change. Nevertheless, the concept of diffusion has today vanished from the archaeological lexicon of British prehistory, and it is unacknowledged as a subject of theoretical speculation, prompting a situation, which can be characterised by a degree of ambiguity (Brami and Garcia Rovira forthcoming) as at the level of interpretation, such arguments continue to be made.

This situation is not unique to the context of the transition but can also be found in other interpretive frameworks. Other more subtle instances can be observed in contexts such as that of the traditional relationship drawn between the Orkney Islands and the Boyne Valley (Ireland) during the later Neolithic. For instance, whilst demonstrating the relationship between the scale of monumentality and the degree of social complexity, Renfrew (1979b) argued that, on the basis of parallels in material culture, 'it was perfectly possible' to suggest the existence of contacts between both regions (Renfrew 1979b, 210). Similarly, in his discussion of burial practices in the Orcadian Neolithic, Sharples (1985) noted that 'the appearance of features with close parallels to Newgrange in Ireland merely underlines a link between the two areas' (Sharples 1985, 71) (see Garcia Rovira 2011 for further discussion). Despite the fact that neither Sharples nor Renfrew promoted the restitution of discourses on diffusion in their interpretive frameworks, they both implicitly posited that cultural encounters can bring about what in archaeology we define as the transfer of cultural traits amongst different social groups.

This ambiguous situation is representative of much archaeological interpretation postdating culture-history. Diffusion - alongside other social phenomena triggered in contexts of intercultural interaction - has remained poorly understood in recent archaeological thought. Most importantly for the present discussion, this situation has been somewhat aggravated in the context of post-processual discourse as, in many instances, the themes explored are characteristically centred in the study of phenomena taking place within specific social-cultural contexts. Agency theory (e.g. Dobres and Robb 2000), contextual archaeology (e.g. Hodder 1982b, 1987) or Barrett's archaeologies of inhabitation (e.g. 1988, 2000, 2001) can serve to illustrate the statement above. However, in this occasion, I will limit the discussion to agency theory, elucidating the inherent constraints placed by the application of theoretical insights centred in the description of phenomena triggered within specific social systems in the context of the British Mesolithic-Neolithic transition. This situation can be identified by taking into consideration the implicit use of diffusion in narratives, which seek to explore agency in this process of change.

Even though the focus of discussion here is centred on the inadequacy of practice theory as a conceptual strategy for grasping phenomena taking place in contexts of intercultural interaction, it is interesting to begin by revealing the ways in which processes of diffusion implied in narratives of the transition have been portrayed when treating agency as intentionality. In this case, whether unidirectional or acting in both directions, diffusion is depicted as a process in which social groups actively acquire alien cultural traits. Although an element of intentionality should not be ignored in this context of change, the outcome offered by these approaches proves problematic in a number of instances. Firstly, it demands a view of culture as an amalgamation of cultural traits, and secondly, it disregards the hermeneutical mechanisms through which new cultural practices are absorbed and transformed within specific social systems over time. Within this paradigm, diffusion is taken as a simple 'transfer of traits from one culture to another' (Renfrew and Bahn 2005, 75), and given the general unidirectionality of most interpretive narratives on the transition, it permeates social evolutionary insights, which, at the level of theory, have been overtly challenged (Borić 2005, 82).

As discussed above, understandings of agency stemming from practice theory have triggered the development of interpretive scenarios which consider how new cultural practices are translated within specific socio-historical settings. This theoretical movement has only partly fed narratives of the transition. In order to understand this situation, it is necessary to briefly focus on the particularity extant in the bodies of theory from which agency theory developed in archaeology. Rather than presenting a comprehensive account of the work of Bourdieu, Giddens, and by extension Heidegger, a task that was done elsewhere (e.g. Barrett 1988, 2000; Dobres and Robb 2000; Jones 1997; Thomas 1996, 2006), this account only seeks to demonstrate the context from which these approaches operate.

Succinctly, the main purpose of Bourdieu's theory of practice (1977, 1984) and Giddens structuration theory (1979, 1984) was to overcome social doctrines which had either subjected human action to the determinism of the social structure, or that had considered it to proceed free of social constraints of any kind. Leaving aside their particularities, these approaches sought to outline the 'dialectical relationship between "agent," a bounded but not determined individual who can alter structures through practice (or praxis), and "structure," the larger, more perduring settings and conditions that result from the ongoing relationships between individuals' (Dornan 2002, 305). To do so, they focused their accounts within the limits of worldly experience, understanding the latter as a meaningful structure of mutually referring elements (Heidegger 1962).

What is important to emphasise is that, whilst the aim of these approaches was to reveal the ways in which social dynamics took place within specific social systems by placing particular emphasis on praxis, the archaeological community has often used these approaches to discuss social dynamics even in instances where change/

transformation took place in interstitial contexts. One such scenario can be found in relation to the ways in which the British Mesolithic-Neolithic transition has been theorised in recent years. On the one hand, it has been noted that the uses of agency within this context can be compared to situations of active intentionality. On the other, the influence of 'practice theory' can be observed in the way in which archaeologists have discussed the mechanisms by which the transition cannot be simply discussed in terms of acquisition, but of transformation and transliteration. Whilst this approach supersedes the naïveté of the former, it implies that transformation only occurs from within specific socio-cultural milieus.

This situation implicitly prompts two basic consequences. Above all, failing to develop theoretical resources that take into consideration the spaces of cultural production engendered in contexts of intercultural interaction, we run the danger of developing narratives in which social dynamics are portrayed as organically constituted, presupposing an understanding of cultural phenomena in terms of purity (see Young 1995). In turn, this situation does not only constraint the development of interpretive narratives centred on the definition of contexts of cultural encounters, but it also prompts the development of bold assertions about the historical contingencies that took place in this context of change. In the light of this situation it appears necessary to develop better understandings on phenomena taking place in interstitial contexts. Whilst a comprehensive account of the ways in which this movement should be pursued is beyond the limits of this account, I seek to draw attention to the spaces of cultural production engendered in spaces 'in-between' (Bhabha 1994) in the next section.

Productive spaces: 'in-between'.

'Hybridity thinking also concerns existing [...] and thus involves different ways of looking at historical and existing cultural and institutional arrangements. [This] suggests not only that things are no longer the way they used to be but were never really the way they use to be, or used to be viewed (emphasis added)'. (Nederveen Pieterse 2009, 97).

Up to this point, it has been argued that despite the benefits of applying insights deriving from practice theory in the context of the transition, these are not sufficient to shed light into the complexities of this context of change as they are constrained by the definition of phenomena taking place within specific social systems. Instead, it has been argued that the histories of change currently sought in the transition can only begin to be explored by developing theoretical resources focused on the kind of phenomena triggered in interstitial contexts. In order to illustrate the benefits of this reorientation, in this section, I aim to shed light into the productive character of the space 'in-between'. However, it is not my intention to replace these bodies of thought with novel theoretical resources. Instead, I would like to suggest that any approach that aims to define phenomena taking place in contexts of intercultural interaction needs to take into consideration the experience in-the-world (Heidegger 1962) as an a priori condition triggering the transformative nature of the 'in-between' (Bhabha 1994). The latter statement can be elucidated by developing an account, which begins in the context of 'hermeneutical philosophy' and moves to the social sphere through insights developed in the fields of post-colonial theory and globalisation studies.

As was noted in the previous section, Heidegger (1962) limited his account to the definition of the common structures of worlds (Brandom 2005, 216); an approach which served as the grounds for practice theory to exhibit the character of internal social dynamics. However, whilst Heidegger limited his ontology to the definition of experience in the world (see 1962), Gadamer (1975) moved to consider what occurs when different worlds – historical horizons - are confronted. He did so, focusing his examination on the experience of reading historical texts; however, his aim was to exhibit hermeneutical mechanisms, which surpassed the limits of textual interpretation (Catoggio 2008).

Since comprehensive accounts of late hermeneutics have been published elsewhere in relation to other issues in archaeology (e.g. Johnson and Olsen 1992; Thomas 2004), on this occasion, I will limit the discussion to the definition of two central elements contained in what Gadamer defined as the 'fusion of horizons' (1975). Departing from a view in which prejudices are taken as preconditions of understanding, Gadamer indicated that understanding a historical text does not mean to 'assimilate the historical other within its own horizon' (Davey 2006, 7), or 'to become fully immersed in the others 'form of life'' (Davey 2006, 7) Contrarily, 'understanding requires and perpetuates a mode of differentiation which sustains understanding as an enduring task' (Davey 2006, 5), bringing forward two fundamental processes.

Gadamer noted that in the event of interpretation, not only an understanding of the *other* emerges but also a 'level of self-awareness' takes place (Lawn 2006, 65). A similar situation is portrayed by Bourdieu (1977) in relation to situations of cross-cultural encounter. As he notes, cross-cultural interaction works as:

'[...] the deliberate methodological suspension of native adherence to the world. The critique which brings the undiscussed into discussion, the unformulated into formulation [...] destroys self-evidence practically'. (Bourdieu 1977, 166).

The notion 'self-reflection' defined by Bourdieu has been taken as a central element to consider processes of identity formation (see Bentley 1987; Jones 1997); a process which brings transformation as it is situational and context dependent. It is also one in which only certain cultural

practices from the broader spectrum of possibilities are associated with the identity of particular groups (Bentley 1987; Jones 1997). In this sense, it is no longer possible to consider social dynamics triggered exclusively within specific social systems. However, this situation is explored by Gadamer (1975), who notes that whilst the self-reflective activity triggered in experiencing alterity is in itself transformative, it is central for the definition of a 'common language' (Gadamer 1989, 388). As he noted:

'To reach understanding in a dialogue is not merely a matter of putting oneself forward and successfully asserting one's point of view, but being transformed into a communion in which we do not remain what we were. [...] A common language can only emerge or be 'worked out' in conversation'. (Gadamer 1989, 378-9).

This insight is fundamental, as it implies that not only does the meeting of alterity trigger the dynamics of tradition, but also alters the latter through the integration of elements which are not taken from the other horizon, but which are constructed through the experience of 'hermeneutical conversations' (Wachterhauser 1986, 202). Similar insights to the ones presented by Gadamer were translated to the social domain in the context of hybridity studies, insights that have been largely explored in the work of Bhabha (1994).

Focused on the deconstruction of the relation traditionally defined between the coloniser and the colonised, Bhabha (1994) draws attention to the centrality of the interstitial contexts as contexts of cultural production. In doing so, he was able to demonstrate that the spaces 'in-between' are productive spaces; interrogative and enunciative (Bhabha 1994) spaces where boundaries 'dissolve'. Therefore, rather than being a place where mixing happens, it is essentially hybrid. It is a space in which 'elements of diverse origin encounter each other and mutual transformation occurs' (Hussain 2005, 12). Following from this, it can be suggested that, intercultural contact is not only transformative in that it triggers a movement of 'self-reflection' but in that it also incorporates cultural elements which are in their essence hybrid.

Whilst a lot more could be said here, I will limit this discussion to exhibit the benefits that are given by placing diffusion within a context that considers the spaces of cultural production engendered in situations of intercultural interaction, and the possibilities this affords for describing the historical contingencies that took place in the context of the transition.

In the first section of this article, I established that current narratives of the transition have posited an implicit understanding of diffusion as a process in which cultural elements are transposed from one group to another, only taking into consideration the creative forces triggered within specific social systems. In these accounts, the contexts in which diffusion occurs are limited to the encounter of autonomous cultural units or social groups, disregarding the transformative nature of interstitial contexts. However, taking into consideration the insights presented in this section, diffusion can instead be characterised as the outcome of the hybridising effects which result from intercultural interaction, and can only be defined insofar as they are observable, be it by the parties involved or by third parties at the moment it occurs, or, as in the case of archaeology, through its definition within temporal and spatial values. This approach is revealing as it triggers new interpretive questions.

If diffusion is no longer considered either as the simple transposition of cultural phenomena or the spread of cultural elements, it is worth enquiring whether it is possible to define the transition as the spread of 'Neolithic things and practices' (Whittle et al. 2011). Whilst in this section, I have posited the possibility to explore contexts of diffusion as a particular situation taking place in hybrid spaces, it is necessary to give explanation for the directionality observed in the 'spread' of the Neolithic in the British Isles; a question which is not resolved in this account but, nevertheless, problematised in the next section.

Moreover, in stating that processes of identity formation are established through the selection of particular cultural practices, and can incorporate elements, which can only be conceived in resulting from interstitial contexts, then it is important to further problematise the established Mesolithic-Neolithic dichotomy. This positioning has begun to crystallise thanks to the Borić's (1985) denunciation regarding the straightforward relationship traditionally given to identity and subsistence regimes in the context of the transition.

Finally, this theoretical reorientation provides us with a series of understandings that are central if we are willing to consider the historical contingencies taking place in this moment of change. With regards to the latter, it is perhaps no longer appropriate to posit scenarios of colonisation or indigenous adaption, establishing a straight forward correlation between the frequencies in which 'Neolithic practices' begin to emerge in different regions of the British Isles at the end of the 5th millennium BC. Post-colonial theory offers new ways of looking at processes of colonisation which move beyond the understanding of the colonised as 'passive victim of colonisation' (Fahlander 2007, 17). Instead, the relationship between the coloniser and the colonised is better understood as an 'interactive, dialogic, two-way, phenomenon' (Trivedi 1993, 15 as quoted in Gandhi 1998, 125) rather than 'the mutual contagion of subtle intimacies' (Appadurai 1990, 38). In which ways new interpretive scenarios would begin to emerge in the context of the transition if theoretical activity placed a thorough movement of theorisation of the spaces 'in-between'?

A matter of reflection

'For Marx, real knowledge of the history of capitalist societies is not to be reduced to theory, though without theory such knowledge is impossible'. (Godelier 1972, 274).

In the last section, I argued that the character of diffusion can begin to be understood by considering it as a particular outcome of the hybridising effects triggered in interstitial contexts. Moreover, I noted that whilst the historical contingencies recently sought in the context of the transition would be benefited by taking into consideration the approach outlined in the last section, a question should be resolved: how can we explain the directionality observed in the British Mesolithic-Neolithic transition?

Moving away from the definition of evolutionary accounts, Hodder (2012) has recently explored a similar question in relation to the spread of farming societies in the Middle East, considering this process to be the result of deep processes of entanglement. Rather than thinking about the emergence of a the Neolithic as a 'package', the elements traditionally identified as 'Neolithic things and practices' (Whittle et al. 2011) not only became tied to each other but also helped guiding change in a specific direction.

The directionality given by processes of entanglement cannot be simply defined in the relationship between things and humans and things and thing but also in the relationships established by humans and humans. New practices would have driven changing social relationships not only within specific social groups but also amongst different social groups. Following from this, it could be suggested that the processes of diffusion observed in the context of the British Mesolithic-Neolithic transition may correspond to long term processes of entrapment (Hodder 2012). This suggestion triggers the development of further questions: how can we reconcile the insights on hybridity developed in the last section and focus on the exploration of the micro-scale with the notions of long temporal processes of entanglement defined by the British archaeologist?

This question can only be complicated further when taking into consideration the changes induced by, using Marxist terminology, new modes of production. Far from inducing to an economic determinist understanding of the happenings of the transition, it is important to consider the ways in which notions on cultural hybridity and the formation of new identities can be reconciled with the idea that new modes of production would have triggered new social structures and, therefore, new social understandings. Would it be possible to develop a multidimensional approach, which triggers new interpretive avenues for the transition? Whilst, as Thomas (2008) indicates, understandings on the transition require a movement of theorisation, it is fair to say that the transition also poses new theoretical puzzles which are yet be resolved.

Bibliography

Ammerman A.J. and Cavalli-Sforza L.L. 1984. *The Neolithic Transition and the Genetics of Populations in Europe*. Princeton, New Jersey, Princeton University Press.

Appadurai A. 1990. Disjuncture and difference in the global culture economy. *Theory, Culture, and Society* 7, 295-310.

Barrett J.C. 1988. Fields of Discourse: Reconstituting a Social Archaeology. *Critique of Anthropology* 7, 5-16.

Barrett J.C. 2000. A Thesis on Agency. In M.A. Dobres, and J.E. Robb (eds.) *Agency in Archaeology*. London, Routledge.

Barrett J.C. 2001. Agency, the duality of structure, and the problem of the archaeological record. In I. Hodder (ed.) *Archaeological Theory Today*. Cambridge, Polity Press.

Bentley C.G. 1987. Ethnicity and Practice. *Comparative Studies in Society and History* 29, 24 -55.

Bhabha H. 1994. *The Location of Culture*. London, Routledge.

Boas F. 1896. The Limitation of the Comparative Method of Anthropology. *Science* 4, 901- 908.

Borić D. 2005. Fuzzy horizons of change: Orientalism and the frontier model in the Mesolithic-Neolithic transition. In N. Milner and P. C. Woodman (eds.) *Mesolithic studies at the beginning of the 21st century*, 81-105. Oxford, Oxbow Books.

Bourdieu P. 1977. *Outline of a theory of practice*. Cambridge, Cambridge University Press.

Bourdieu P. 1990. *Reproduction in education, society and culture*. London, Sage.

Brami M., Garcia Rovira I. (In Press) *Diffusion beyond diffusionism*.

Brandom R. 2005. Heidegger's Categories in Being and Time. *The Monist* 66(3) 387-409.

Wrathall, M. 1992 (ed.) *A Companion to Heidegger*. Oxford, Blackwell.

Case H.J. 1969. Neolithic Explanations. *Antiquity* 43, 176-86.

Childe V.G. 1925. The Dawn of European Civilisation. London, Kegan, Trench, Trubner and Co.

Cummings V. 2007. From Midden to Megalith? The Mesolithic-Neolithic transition in Western Britain. *Proceedings of the British Academy* 144, 493-510.

Cumming V. and Harris O. 2011. Animals, people and places: the continuity of hunting and gathering practices across the Mesolithic-Neolithic transition in Britain. *European Journal of Archaeology* 14(3), 361- 382.

Davey N. 2006. *Unquiet Understanding: Gadamer's Philosophical Hermeneutics*. Albany, NY, State University of New York Press.

Dobres M.A. and Robb J.E. 2000. *Agency in Archaeology*. London, Routledge.

Dornan J.L. 2002. Agency and Archaeology: Past, Present, and Future. *Journal of Archaeological Method and Theory* 9(4), 303-329.

Fahlander F. 2007 Third Space Encounters: Hybridity, Mimicry and Interstitial Practice. In P. Cornell and F. Fahlander (eds.) *Encounters, Materialities, Confrontations. Archaeologies of Social Space and Interaction*. Cambridge, Cambridge Scholar Publishing.

Fewster K.J. 2001. Petso's Field: ethnoarchaeology and agency. In K. Fewster, and M. Zvelebil (eds.) *Ethnoarchaeology and hunter-gatherers. Pictures at an exhibition*. Oxford, British Archaeological Reports.

Gagamer H.G. 1975. *Truth and Method*. London, Shed and Ward.

Gadamer H.G. 1989. *Truth and Method*. New York, Crossroad.

Gandhi L. 1998. *Postcolonial theory: a critical introduction*. Edinburgh, Edinburgh University Place.

Garcia Rovira I. 2011. Re-thinking diffusion 'in-between'. *Cultural encounters, time and the formation of hybrid identities*. Unpublished Ph.D thesis, University of Manchester.

Garrow D. and Sturt F. 2011. Grey waters bright with Neolithic Argonauts? Maritime connections and the Mesolithic-Neolithic transition within the 'western seaways' of Britain, c.5000-3500 BC. *Antiquity* 85, 59-72.

Giddens A. 1979. *Central problems on social theory: action, structure and contradiction in social analysis*. London, Macmillan.

Giddens A. 1984. *The constitution of society: outline of the theory of structuration*. Cambridge, Polity.

Gilroy P. 1993. *The Black Atlantic: Modernity and Double Consciousness*. Cambridge MA, Harvard University Press.

Heidegger M. 1962. *Being and Time*. New York, Harper and Row.

Hodder I. 1982a. *Symbols in action: ethnoarchaeological studies of material culture*. Cambridge, Cambridge University Press.

Hodder I. (987b. *The Archaeology of Contextual Meanings*. Cambridge, Cambridge University Press.

Hodder I. 2012. *Entangled. An Archaeology of the relationships between humans and things*. Oxford, Willey-Blackwell.

Hussain Y. 2005. *Writing diaspora: South Asian women, culture, and ethnicity*. Aldershot, Ashgate Publishing Limited.

Johnsen H., Olsen B. 1992. Hermeneutics and archaeology: on the philosophy of contextual archaeology. *American Antiquity* 57(3), 419-436.

Jones S. 1997. *The archaeology of ethnicity: constructing identities in the past and present*. London, Routledge.

Lawn C. 2006. *Gadamer: a guide for the perplexed*. London, Continuum.

Milner N. Craig O.E., Bailey G.N., Pederson K. and Anderson S.H. 2003. Something fishy in the Neolithic? A re-evaluation of stable isotope analysis of Mesolithic and Neolithic coastal populations. *Antiquity* 78, 9-22.

Montelius O. 1905. *Orienten och Europa*. Stockholm, Antiqvarisk Tidskrift för Sverige, B XIII.

Olsen B.J. 2001. The end of history? Archaeology and the politics of identity in a globalised world. In R. Layton,

P. Stone, and J. Thomas (eds.), *The Destruction and Conservation of Cultural Property*. London, Routledge.

Piggott S. 1954. *The Neolithic cultures of the British Isles: a study of the stone-using agricultural communities of Britain in the second millennium BC*. Cambridge, Cambridge University Press.

Renfrew C. 1979. *Investigations in Orkney*. London, Society of Antiquaries of London, Thames and Hutson.

Richards M.P. and Hedges R.E.M. 1999. Stable isotope evidence for similarities in the types of marine foods used by late Mesolithic populations on the Atlantic coast of Europe. *Journal of Archaeological Science* 26, 217-222.

Rushdie S. 1982. *Imaginary Homelands*. London, Review of Books, 7-20.

Richards, M.P., Schulting, R.J., and Hedges, R.E.M. 2003. Sharp shift in diet at onset of Neolithic. *Nature* 425, 366.

Schulting R.J., Richards M.P. 2002. Finding the coastal Mesolithic in southwest Britain: AMS dates and stable isotope results on human remains from Caldey Island, Pembrokeshire, South Wales. *Antiquity* 76, 1011-1025.

Sharples N. 1985, Individual and community: the changing role of megaliths in the Orcadian Neolithic. *Proceedings of the Prehistoric Society* 51, 59-74.

Sheridan J.A. 2003. French connections I: spreading the marmites thinly. In I. Armit, F. Murphy, E. Nelis and D.D.A Simpson (eds.), Neolithic settlement in Ireland and western Britain, 3-17. Oxford, Oxbow.

Stross B. 1999. The Hybrid Metaphor: from biology to culture. *The Journal of American Folklore* 112, 254 – 267.

Sheridan J.A. 2004. Neolithic connections along and across the Irish Sea. In V. Cummings, and C. Fowler (eds.), *The Neolithic of the Irish Sea: materiality and traditions of practice* 9-21. Oxford, Oxbow.

Thomas J.S. 1988. Neolithic explanations revisited: the Meoslithic-Neolithic transition in Britain and South Scandinavia. *Proceedings of the Prehistoric Society* 54, 59-66.

Thomas J.S. 1996. *Time, culture and identity: an interpretive archaeology*. London, Routledge.

Thomas J.S. 2003. Thoughts on the `repacked? Neolithic revolution. *Antiquity* 77, 67-74.

Thomas J.S 2004. The great dark book: archaeology, experience and interpretation. In J. Bintliff (ed.) *A Companion to Archaeology* 21-36. Oxford, Blackwell.

Thomas J.S. 2006. Gene-flows and social processes. The potential of genetics and archaeology. *Documenta praehistorica* 33, 51-59.

Thomas J.S. 2008. The Mesolithic-Neolithic transition in Britain. In J. Pollard (ed.) *Prehistoric Britain* 58-89. Oxford, Blackwell.

Tilley C. 1994. *A Phenomenology of Landscape: places, paths and monuments*. Oxford, Berg.

Tilley C. 2007. The Neolithic sensory revolution: monumentality and the experience of landscape. *Proceedings of the British Academy* 144, 329-334

Trivedi H. 1993. *Colonial Transactions. English literature and India*. Calcutta, Papyrus.

Wachsterhauser B.R. 1986. Must we be what we say? Gadamer on truth in the social sciences. In B.R. Wachterhauser (ed.) *Hermeneutics and modern philosophy*. Albay, University of New York Press.

Warren G. 2007. Mesolithic myths. *Proceedings of the British Academy* 144, 311-328.

Whittle A. 2003. *The archaeology of people: dimensions of Neolithic life*. London, Routledge.

Whittle A. 2007. Going over: people and their times. *Proceedings of the British Academy* 144, 617- 628.

Whittle A., Healy F. and Bayliss A. 2011. *Gathering time. Dating the Early Neolithic enclosures of Southern Britain and Ireland*. Oxford, Oxbow.

Young R.J.C. 1995. *Colonial Desire: hybridity in theory, culture and race*. London, Routledge.

CHAPTER 5
BRAVE NEW WORLD,
THE PATHS TOWARDS A NEOLITHIC SOCIETY IN SOUTHERN
SCANDINAVIA

Mats Larsson

Linnaeus University

Abstract: *This article is about the neolithisation in Southern Scandinavia. The background in the late Mesolithic Ertebølleculture is discussed and things like import items, cemeteries and settlement pattern are discussed. Climatic change and the resulting change in the procurement strategy is seen as important factors in the obviously rapid change during the late Mesolithic. The Early Neolithic in Scania is then discussed in some length and issues like chronology and settlement pattern are discussed. One of the main issued is that there is not that marked change in settlement pattern. It has often been proposed that the inland was a more or less unknown entity but this was not the case. Early Neolithic man settled though in very distinct areas with sandy soils and proximity to water, thus being able to use a wide spectrum of ecological niches.*

Keywords: *Late Mesolithic, Early Neolithic, settlement pattern, pits, houses*

The Late Mesolithic. A period of change

From 5500 BC until 4000 BC the Ertebølleculture existed in the whole of southern Scandinavia with concentrations to Denmark and Southern Sweden. The Ertebølleculture is primarily known for maybe three things; the introduction of pottery, the vast kitchenmiddens and the large cemeteries. In the period 4300-4000 BC, in what Klassen (2004, 103) calls import phase 5, there is a marked increase in the number of what may be termed prestigious objects like T –shaped antler axes and Shoe last axes. These foreign items we can perceive were loaded with potent images and had a symbolic value. They played an active part in the transformation of society (Fischer 2002). As early as c 4600 we also have evidence from northern Germany (Grube –Roenhof) of domesticated cattle. This can be seen as an import with no traceable influence on the economy (Hartz, S, Lübke, H and Terberger, T. 2007, 587)

There is also evidence for a change in the use of marine resources during the period 4800-4600 BC. It is obvious that there is a rise in the use of marine resources like oysters and at the same period of time we see a marked growth of the middens. This culminates in the period 4600-4000 BC (Andersen 1993, 74). The kitchenmiddens on Jutland are due to their complex stratigraphy of great importance in the interpretation of the settlement history and chronology of the late Mesolithic as well as the Earliest Neolithic. We can also, based on the middens as well as other sites, study the diversity of environment used at the time. A small site like Agernœs on the northwest corner of the island of Funen might typify a smaller specialized late Ertebølle site (Richter and Noe-Nygaard 2004). The site is dated to

4200 BC contain 32 different animal species. Evidence from excavations in the middens during the last decades has also revealed that these were divided into smaller sites with hearths, work floors etc. (Andersen 2001, 32). The notion that the late Ertebølle sites were very large is in other words under discussion. They were continuously used for a long period of time though. In this context it is of interest to have a closer look at one special midden. This is Björnsholm in the central Limfjord area in northern Jutland (Andersen 1993, 59). Close to it, an early Neolithic grave with a timber construction was excavated (Andersen and Johansen 1992). The Ertebølle layers date to 5050 - 4050 BC and the Early Neolithic site is dated to between 3960 and 3520 BC (Andersen 1993, 61). This implies a more or less continuous habitation for about 1000 years. This is maybe en extreme, as we also know of sites like Meilgård that was in use for about 500 years (Liversage 1992, 102). An area behind the midden at Björnsholm was excavated as well. Only very ambiguous traces of habitation were revealed (Andersen 1993, 66; 2001, 26-29). The only more substantial features found in the midden are hearths and these occur in all levels of the midden. There is however no traces of huts, pits or postholes so it is still an enigma where people actually lived. The answer to this problem might be that the middens were a type of coastal sites at which people went about doing their daily routines, fishing, hunting and so forth. They did not actually live here though. We could maybe also see these locations as the places for communal feasting used at particular times of the year. But the question still lingers; where did people actually live. It might be that the middens were only a part of the actual settlement area but this notion doesn't really answer the question.

What we can say however is that the people of the Ertebølleculture used a diversity of ecological areas ranging from the fjords of eastern Jutland to the inland of Scania. An inland site like Ringkloster from the late Ertebølleculture on Jutland is in this context of great importance (Andersen 1975).

The cemeteries that in some cases are associated with the settlement sites are still quite few. The first cemetery from the period that was excavated is Vedbæk north of Copenhagen on Zealand. This is rather small with 22 individuals of both sexes and of various ages. It is dated to c 5000 BC (Albrethsen and Brinch-Petersen 1977). The largest of the known cemeteries is Skateholm 1 with more than 60 burials (Larsson, L. 1984).

The apparent lack of more substantial traces of huts or houses, especially from the later part of the Ertebølleculture, has been one of the most discussed issues in Scandinavian Mesolithic research over the years; why do we not find any more substantial traces of occupation (for a review see Biwall *et al.*1997; Cronberg 2001, 81-89). One of the most intriguing new aspects of the late Mesolithic has therefore been the excavation of houses and huts at the site Tågerup in western Scania (Cronberg 2001, 89-149). The site is not located directly at the coast but in a sheltered position in a deep lagoon. The oldest remains are from the Kongemose culture (Karsten and Knarrström 2001). One of the houses, House I, was a circular construction rebuilt in two stages. The other one, House II, was an 85 m2 rectangular longhouse with a singular row of roof supporting posts. House III has been described as a shelter.

The chronology of the houses is complicated as discussed by Cronberg (2001, 147). The radiocarbon dates are, except for two, all way to young and are mostly Neolithic or Bronze Age. The dating of the houses therefore had to be based on a study of the transverse arrowheads (Vang-Petersen 1984). This dates the houses to the early and middle phases of the Ertebölle culture. Interestingly enough a couple of pottery sherds are mentioned from House III. They are accordingly dated to the Ertebölle culture (Cronberg 2001, 143). Pottery is not adopted in Denmark and Scania until c 4600 BC (Hallgren 2004, 136). If the interpretation of House III is correct, it is not older than 4600 BC. The house structures from Tågerup might imply that people during the late Mesolithic lived in large communal houses.

Man, memory and landscape

The Neolithisation in Scania

I will in the following argue for that at the start of the Early Neolithic there is a change of focus in the settlement location and connected to this also a change in how people perceived the landscape that they lived in. The focus will be on southernmost Sweden, Scania. The starting point will be the often discussed change in settlement pattern between the late Mesolithic and the early Neolithic (Larsson M.

1984). The large coastal Ertebølle sites were more or less abandoned as permanent settlements. They were however still used seasonally by the early farmers. New areas inland were occupied, preferably sandy soils close to water. The sites have usually been seen as small, about 600-1000m2 but this notion is however under discussion today (Larsson M. 1984; Andersson 2003, 175).

We can tell a story of how people at this time re-invented themselves and their relationship with both other people and the surrounding landscape. This view goes well with the one that Carlstein (1982, 39) sees as "a web of individual-paths in time-space". This notion can also in many ways be compared to what Tim Ingold (2000, 193) in a discussion of temporality and landscape has written, "In short, the landscape is the world as it is known to those who dwell therein, who inhabit its places and journey along the paths connecting them". This he calls a "taskscape" The landscape is thus both agency and time embodied (Ingold 2000).

Most agency theorists agree on that agency is a socially significant quality of action rather than being synonymous with action itself (Dobres and Robb 2000, 8). It is also important to stress, as John Barrett (2000, 61) has stated, that action, time, space and agency work together and carry the past into the future. In using agency theory we can investigate how cultural meanings and social structures are constructed and transformed through peoples interactions with others, loosely termed social practice (Jordan 2004, 112)

One main issue that has been discussed is to what degree Neolithic man changed his concept of land and the myths and stories about it. As John Barrett (1994, 93) has suggested this might have a long history perhaps going back to the Mesolithic in that those places might have been part of a much wider seasonal cycle of movement. This is also what the evidence from the middens suggests. It is obvious that they were seasonally used. Some for the hunting of swans and migrating birds like the site Aggersund (Andersen 1978). Some of the middens like Norsminde, Ertebølle and Björnsholm have traces of earlier occupation. These pre-midden layers are radiocarbon dated to between 4960-4600 cal BC (Andersen 2001, 24). Places like theses could be termed "persistent places" and such places would have engendered a sense of time and belonging (Cummings 2002, 79).

What actually motivated this shift in settlement and why did people move? Was the shift actually that radical? As discussed above the inland area was by no means an unknown entity. The change in settlement area might be seen as a functional response to economic changes but also to climatic change. There is a marked increase in temperature at the Atlantic-subboreal transition (Karlén and Kuylenstierna 1996) as well as dramatic changes in the shoreline displacement. This is especially noteworthy in eastern Denmark where the shoreline displacement caused marked changes in the landscape (Larsson L.2007, 605). Taken together these things altered the landscape in the long run and an increase in temperature also facilitated the growing of cereals. Changes in the ecology changed and altered the structural conditions under which the new subsistence could operate. Mesolithic man of course had knowledge of different ecological niches and that is why the move might not have been that upsetting. Motivations for doing this, we can argue, must have been formulated into strategies by people who had a certain level of knowledge about their social and natural environment. People were active agents in the way in which the settlement sites, the farming plots and so on were chosen. They created meaningful landscapes, which helped them to develop a sense of group identity as well as a self-identity. People also mixed their material culture bringing in both new items while still preserving some old ones. I would here draw the attention to the obvious similarity between the flint inventory of the Ertebölle culture and the earliest Funnel Beaker culture (Oxie group). The existence of transverse arrowheads, flake axes and singular core axes in the latter has been noted and discussed as evidence for a close connection between these. This close connection actually only exists in the earliest Neolithic (Oxie/Svenstorp) and is obviously gone in the later stages (Bellevuegården/Virum) (cf Larsson M.1984, 162; Fischer 2002, 351-355). The term "creolisation" might be useful in this context. The term refers to a process whereby men and women actively blend together elements of different cultures to create a new culture. Creolisation is perceived as a more active process and one that involves, by definition, a give and take between peoples of diverse cultural traditions (Cohen and Toninato 2010, 1). The term leads Zvelebil to formulate the 'Neolithic creolisation hypothesis' based on language, migration and the merging of different cultures (Zvelebil 1995). This is why the concept of agency is important as it recognize that men and women make well-supported choices, and take action to realize these choices (Dobres and Robb 2000, 133). In the earliest Neolithic individuals or groups of people who had a certain level of knowledge about their social and natural environment must have formulated individual and collective motivations - reasons and justifications for doing things - into strategies.

As mentioned above Mesolithic man had previously both moved and stayed in the inland of Scania as is well known from several settlement sites like for example Bökeberg (Karsten 2001). As noted by Magnus Andersson (2003, 161) though in his work in western Scania there are few if any traces of Mesolithic habitation on the Early Neolithic inland settlements. There is obviously a different strategy at work here; the earliest farmers chose quite other areas for settlement. I we look at the distribution of the earliest polished flint axes, the pointed butted variant; it is obvious that the inland took on a whole new dimension (Hernek 1989; Jennbert 1984). It is now time to take a closer look at the settlement sites.

Several of the earliest Neolithic settlements had a distinct location in the landscape. They were situated on ridges or small hills in the undulating landscape. People located their settlements to areas with a large ecological diversity. The sites are close to water and in general located on sandy soils. This is especially true for the sites with large amounts of pits like Svenstorp and Månasken in SW Scania (Larsson M. 1984; 1985). The pits are often layered meaning that they were actually recut and reused. Large amounts of flints debris are found in the pits, but also obviously unused implements like flake axes, flake scrapers and in some cases even complete axes and vessels (Larsson M.1984; 1985). The interpretation of these pits has usually been very functional; they were waste pits. Some of the pits with complete axes or vessels have been re-interpreted though as ritual pits (Karsten 1994; Andersson 2003, 169; Rogius et al.2003; Larsson M. 2007).

Above I mentioned that memory is important in creating new identities but also in recollecting the distant past. The coastal sites were reoccupied and used seasonally during the earliest Neolithic period thus creating links with past generations. Strassburg (2000, 292) mentions that there is a marked change in funerary rites 4900-4700 cal BC as for example observed at the Skateholm cemetery. People seem to use red-ochre sparingly and there are very few, if no, depositions of red deer antlers. Interestingly enough, if this notion is correct, we can in the earliest Neolithic see that red deer antlers were once again deposited in not graves but in pits that very often are associated with long barrows. At the site Almhov in Malmö (Western Scania) several pits with antlers were excavated a few years back (Gidlöf *et al.* 2006; Rudebeck 2010, 83-253). These pits are associated with three long barrows. In for example pits 6, 232 and 3868 several parts of red deer antlers were found (Gidlöf *et al.* 2006, 64, 65, 71). Pit 6 was radiocarbon dated to 3940-3690 cal BC (Ua-17156) that is very early in the Neolithic sequence (Gidlöf *et al.* 2006, 61). Liliana Janik (2003, 116) discusses the possible function of different animals as markers of cultural identity. In the light of the discussion above, this is a highly intriguing possibility. If the significance of different animals changed over time it might be vital for the interpretation of the ways in which the Early Neolithic unfolded and evolved. The notion of memory is important here as well.

The reuse of the coastal areas for different purposes is also evident from the earlier mentioned site Björnsholm. Here a grave structure with an occupation layer was excavated in 1988. The stone-lined pit was found together with two large pits, a ditch and three intact pots and belongs to a group well known Danish Long-Barrows (Andersen and Johansen 1992, 38). It was situated c 20 m to the rear of the midden. The Long Barrow was built very close to the Ertebölle midden and to the subsequent Early Neolithic settlement. This could also be seen as a way of humans to connect with both the past and the present. The construction of this Long Barrow and its use would also add to other memories associated with this place. It connects the midden as a place of both life and death to the later Early Neolithic monument.

Through narratives going back centuries in which these sites and middens figured meaning and remembrance took on a new dimension and they were transformed into something much more potent in the Early Neolithic.

How then did people adopt to the new situation? As mentioned above there is a change in the settlement pattern at the beginning of the Early Neolithic, c 3950 cal BC. New areas inland are used and the digging of large number of pits on some sites as well as the building of the first Long Barrows is all parts of this new scenario.

So why discard complete vessels and implements? We can, as mentioned above, not just see the pits as waste pits but as evidence for something much more profound. Richard Bradley (2000, 131) has said, referring to Britain that these artefacts were being returned to the elements from which they were formed. According to Julian Thomas (1999) the digging of pits are associated with feasting and the items deposited in the pits are a sort of remembrance of these feasts. This must in many ways have changed how people perceived the landscape. By bringing together different elements, the sites eventually became a microcosm of the landscape as a whole. The performance of rituals was an important part of the structuration of society and they helped people to not only connect and re-connect with the ancestors, but also with the future. This topic has been widely discussed during the last decade or so (Thomas 1991 ch.4; Bradley 2000, 2005). People gave specific places like rocks and streams names and in creating paths in the forest they also created links between both places and people. Myths and stories are then told about it and the place thus becomes historical (Thomas 1996, 89). If we go with this notion it is obvious that monuments as well as other constructions fitted into a landscape already filled with potent and symbolic places (Cummings 2002, 107). But it was now much more important to secure an identity and to make a lasting impact on the landscape.

Thus we can view the first monuments as a medium to preserve social order and it was at the same time part of a communal activity. The building of a monument significantly alters people's roles in the landscape and their view of it. Structures are both the medium and the outcome of social practices (Parker Pearson and Richards 1994, 3). This notion has rather elegantly been expressed by Matthew Johnson (1994, 170) as follows "Landscape is all about a sense of place; architecture is simultaneously a moulding of landscape and the expression of a cultural attitude towards it".

There is one last remark to be made regarding the way in which material culture or symbols helped to constitute a New World. Christopher Gosden (1994, 35) has proposed that standardized material forms provide a support for people when dealing with rapid changes in society. This might thus be seen as a part of the communal memory. Mixing old and new elements made it possible for people to create "a new world" (Thomas 1996, 37). Public meanings

especially of ethnic identity and interpretations were negotiated and contested. In times of rapid change people used, re-invented and re-used both new and old principles while still being able to communicate, as we have seen, through the medium of material culture.

What I in this article have argued for is that there are many similarities between the late Ertebølleculture and the Early Funnel beaker culture in their use of material culture and still the changes that we can observe are profound, new settlement pattern, new economy and the building of the first monuments. The re- use of the middens and the coastal sites, the use of red deer antlers and the similarity in material culture all point to a common history. Implicit knowledge, habitual practice and material culture may be seen as forms of cultural inheritance, which are passed between generations, and modified by innovation. In addressing this process it is important to recognise that artefacts, architecture and landscape are not merely the outcomes of human action, but the media through which human projects are carried forward. Taken together the past helped people not only to connect and re-connect with the ancestors, but also with the future. In this way a "New world" was created. Or was it a "sign of the gods"? (Larsson L. 2007, 607)

References

Albrethsen, S. E. and Petersen, E. B. 1977. Excavations of a Mesolithic cemetery at Vedbæk, Denmark. *Acta Archaeologica* 47, 1-28.

Andersen, S.H. 1975. Ringkloster, en jysk inlandsboplads med Ertebllekultur. *KUML*, 11-94.

Andersen, S.H. 1979. Aggersund. En Ertebølleboplads ved Limfjorden. *KUML*, 7-50.

Andersen, S. H. 1993. Björnsholm. A stratified Kökkenmödding on the Central Limfjord, North Jutland. *Journal of Danish Archaeology* 10, 59-96.

Andersen, S. H. 2001. Danske køkkenmøddinger anno 2000. In O. L. Jensen, S. Sörensen and K. M. Hansen (eds.), *Danmarks jaegerstenalder-stus og perspektiver*, 21-43. Hörsholm, Hörsholms Egns Museum.

Andersen, S. H. and Johansen, E. 1992. An Early Neolithic grave at Björnsholm, North Jutland. *Journal of Danish Archaeology* 9, 38-58.

Andersson, M. 2003. Skapa plats i landskapet. Tidig- och mellanneolitiska samhällen utmed två västskånska dalgångar. Malmö, Acta *Archaeologica Lundensia* 8, 22.

Barrett, J. 1994. *Fragments from Antiquity. Archaeology of Social Life in Britain, 2900-1200 BC*. Oxford, Routledge.

Barrett, J. 2000. A Thesis on Agency. In M-A. Dobres, and J. Robb (eds.), *Agency in Archaeology*, 61-69. London, Routledge.

Biwall, A, R., Hernek, B., Kihlstedt, M., Larsson, M., and Torstensdotter-Åhlin, I. 1997. Stenålderns hyddor och hus i Syd- och Mellansverige. In M. Larsson and E. Olsson (eds.), *Regionalt och interregionalt. Stenåldersundersökningar i Syd och Mellansverige* 265-

300. Stockholm, Riksantikvarieämbetet.Arkeologiska Undersökningar Skrifter nr. 23.

Bradley, R. 2000. *Archaeology of Natural Places.* London, Routledge.

Bradley, R. 2005. *Ritual and Domestic Life in Prehistoric Europe.* London, Routledge.

Carlstein, T. 1983. *Time Resources, Society and Ecology, Preindustrial Societies v. 1* Lund, Allen & Unwin.

Cohen, R. and Toninato, P. 2010. The crelozation debate; analysing mixed identities and cultures. In, R. Cohen. and P. Toninato (eds.), *The creolization reader. Studies in mixed identities and cultures*, 1-23. London, Routledge.

Cronberg, C. 2001. Husesyn. In P. Karsten and B. Knarrström (eds.), *Tågerup specialstudier, Skånska spår-arkeologi längs Västkustbanan*, 82-154. Stockholm, Riksantikvarieämbetet.

Cummings, V. 2002. All cultural things, actual and conceptual monuments in the Neolithic of western Britain. In C. Scarre (ed.), *Monuments and landscape in Atlantic Europe. Perception and society during the Neolithic and Early Bronze Age*, 107-122. London, Routledge.

Dobres, M-A. 2000. *Technology and Social Agency.* London, Routledge.

Dobres, M-A. and Robb J. 2000. Agency in archaeology, paradigm or platitude. In M-A, Dobres and J. Robb (eds.), *Agency in Archaeology*, 3-19. London, Routledge.

Fischer, A. 2002. Food for feasting? An evaluation of explanations of the neolithisation of Denmark and southern Sweden. In A. Fischer and K. Kristiansen (eds.), *The Neolithisation of Denmark*, 343-393. Sheffield, Sheffield University Press.

Gidlöf, K., Hammarstrand-Dehnman K. and Johansson T. (eds.) 2006. *Almhov-delområde 1. Rapport över arkeologisk slutundersökning Rapport nr 39.* Malmö, Malmö Kulturmiljö.

Gosden, C. 1994. *Social being and time.* Oxford, Blackwell.

Hallgren, F. 2004. The introduction of ceramic technology around the Baltic Sea in the 6th Millennium. In H. Knutsson (ed.), *Coast to Coast. Arrival*, 123-143. Uppsala, Coast to coast books no. 10.

Hartz, S., Lübke, H. and Terberger, T. 2007. From fish and seal to sheep and cattle, New research into the process of neolithisation in northern Germany. In, A. Whittle and V. Cummings (eds.), *Going over. The Mesolithic-Neolithic Transition in North-West Europe*, 567-594. Proceeding of the British Academy 144, Oxford University Press.

Hernek,R. 1989. Den spetsnackiga yxan av flinta. *Fornvännen* 83, 216-223.

Ingold, T. 2000. *The perception of the environment, essays on livelihood, dwelling and skill.* London , Routledge.

Janik, L. 2003. Changing paradigms, food as a metaphor for cultural identity. In M. Parker-Pearson (ed.), *Food, Culture and Identity in the Neolithic and Early Bronze Age*, 113-125. Oxford, British Archaeological Reports International Series 1117.

Jennbert, K. 1984. Den produktiva gåvan. *Acta Archaeologica Lundensia* 4, 16.

Johnson, M. 1994. Ordering houses; creating narratives. In, M. Parker-Pearson and C. Richards (ed.), *Architecture and Order. Approaches to social space*, 153-160. London, Routledge.

Johnson, M. 2000. Self- made men and the staging of agency. In M. A. Dobres and J. Robb (ed.), Agency in Archaeology, 213-231. London, Routledge.

Jordan, P. 2004. Examining the Role of Agency in Hunter Gatherer Cultural Transmission. In A. Gardner (ed.), 107-34 *Agency Uncovered, Archaeological Perspectives on Social Agency, Power and Being Human.* London, UCL Press.

Karlén, W. and Kuylenstierna, J. 1996. On solar forcing of Holocene climate evidence from Scandinavia. *The Holocene* 6, 359-65.

Karsten, P. 1994. Att kasta yxan i sjön, en studie över rituell tradition och förändring utifrån skånska neolitiska offerfynd. *Acta Archaeologica Lundensia*, 80, 23.

Karsten, P. 2001. *Dansarna från Bökeberg.* Stockholm, Riksantikvaireämbetet.

Karsten, P. and B. Knarrström (ed.) 2001. *Tågerup specialstudier. Skånska spår-arkeologi längs Västkustbanan.* Stockholm, Riksantikvarieämbetet.

Klassen, L. 2004. *Jade und Kupfer. Untersuchungen zum Neolithisierungsprozess im westlichen Osteseeraum undter besonderes Berucksichtung der Kulturentwicklung Europas 5500-3500 BC.* Aarhus, Jutland Archaeological Society.

Kihlstedt, B., Larsson, M., Nordqvist, B. 1997. Neolitiserngen i Syd-Väst- och Mellansverige-social och ideologisk förändring. In M. Larsson and E. Olsson(eds.), *Regionalt och interregionalt. Stenåldersundersökningar i Syd- och Mellansverige*, 85-133. Stockholm, Riksantikvarieämbetet.Arkeologiska Undersökningar Skrifter nr. 23.

Koch, E. 1998. *Neolithic bog pots from Zealand, Mön, Lolland and Falster Serie B, vol.16.* Copenhagen, Nordiske Fortidsminder.

Larsson, L. 1988. *Ett fångstsamhälle för 7000 år sedan. Boplatser och gravar i Skateholm.* Signum. Kristianstad.

Larsson, L. 2007. Mistrust traditions, consider innovations? The Mesolithic-Neolithic transition in southern Scandinavia In A. Whittle and V. Cummings (eds.), *Going over. The Mesolithic-Neolithic Transition in North-West Europe* 597-617. Proceeding of the British Academy 144, Oxford, Oxford University Press.

Larsson, M. 1984. Tidigneolitikum i Sydvästskåne. Kronologi och bosättningsmönster. *Acta Archaeologica Lundensia* 4, 17.

Larsson, M.1985. *The Early Neolithic Funnel Beaker Culture in South-West Scania, Sweden.* Oxford, British Archaeological Reports International series 264.

Larsson, M. 2007. I was walking through the wood the other day. Man and landscape during the late Mesolithic and early Neolithic in Scania, southern Sweden. In, B. Hårdh, K. Jennbert and D. Olausson (eds.), *On the road. Studies in honour of Lars Larsson*, 212-217. Stockholm, Almqwist and Wiksell International.

Larsson, L. 2007. Mistrust traditions, consider innovations? The Mesolithic-Neolithic transition in southern Scandinavia. In A. Whittle and V. Cummings (eds.), *Going over. The Mesolithic-Neolithic Transition in North-West Europe* 595-616. Proceeding of the British Academy 144, Oxford, Oxford University Press.

Parker-Pearson, M. and Richards, C. 1994. Architecture and order, Spatial representation and archaeology. In M. Parker-Pearson and C. Richards (eds.) *Architecture and Order. Approaches to social space*, 38-73. London, Routledge.

Richter, J. and Noe-Nygaard, N. 2003. A late Mesolithic hunting station at Agernæs, Fyn, Denmark , differentiation and specialisation in the late Ertebølle culture-heralding introduction of agriculture?.*Acta Archaeologica* 74,1-64.

Rogius, K, Eriksson, N. and Wennberg, T. 2003. Buried Refuse? Interpreting Early Neolithic Pits. *Lund Archaeological Review* 20, 7-19.

Rudebeck, E. 2010. I trästodernas skugga- monumentala möten I neolitiseringens tid. In B. Nilsson and E. Rudebeck (eds.), *Arkeologiska och förhistoriska världar. Fält, erfarenheter och stenåldersplatser i sydvästra Skåne*, 83-253. Malmö, Malmö Museer, Arkeologi enheten,

Strassburg, J. 2000. *Shamanic shadows. One hundred generations of undead subversion in southern Scandinavia 7000-4000 BC Stockholm.* Stockholm, Studies in Archaeology 20.

Thomas, J. 1991. *Rethinking the Neolithic.* Cambridge, Cambridge University Press.

Thomas, J. 1996. *Time, Culture and Identity. An Interpretative Archaeology.* London, Routledge.

Thomas, J. 1999. *Understanding the Neolithic.* London, Routledge.

Vang-Petersen, P. 1984. Chronological and regional variation in the late Mesolithic of eastern Denmark. *Journal of Danish Archaeology* 3, 7-19.

Zvelebil, M. 1995. Indo-European origins and the agricultural transition in Europé. *Journal of European Archaeology* 3, 33 – 70.

Zvelebil, M. 2003. People behind the Lithics. Social life and social conditions of Mesolithic communities in temperate Europe. In L. Bevan and J. Moore (eds.), *Peopling the Mesolithic in a Northern Environment*, 1-26. Oxford, British Archaeological Reports International Series 1157.

CHAPTER 6
SHETLAND:
THE BORDER OF FARMING 4000-3000 BC:
PEOPLING AN EMPTY AREA?

Ditlev L. Mahler

The National Museum of Denmark

Abstract: *The paper deals with the demographic questions in connection with the Neolithic expansion to Shetland between 4000 and 3700 BC. There are traces of a Mesolithic population, though, at West Voe, but the size of the population and the duration of the settlement is uncertain. The most likely model for the Neolithic expansion is the so called "pulse model", and it is suggested that there must have been contacts may be back to the mother population, but we have at the moment no archaeological traces of such a communication. The Sarqqaq Culture of Greenland and the Lapita expansion in Oceania around 2000 BC both contains traces of exchange chains, and the exchanged items are seen as "mediators" for managing the demographic problems. We do not know what the mediator could be concerning the Shetland Islands, yet.*

Keywords: *Shetland, Neolithic, Demography, Expansion, Exchange*

Introduction

Through an examination of the various elements constituting the Neolithic Period, this research project will provide a comparative analysis involving the Neolithic communities in Shetland and Scandinavia with particular emphasis on Southern Scandinavia. The comparative elements do not merely involve farming such as grain cultivation and domesticated animals, but also ornamented ceramics, polished tools, the use of the ard, ritual depositions in wetlands as well as monumental structures including megaliths and gathering places. These diverse elements constitute the Neolithic Period in Southern Scandinavia; however, they are not all present simultaneously. Rather, some are present at an early stage whereas others appear relatively quickly after. Renewed analyses point to a rising ritualisation of the early Neolithic communities as part of an explanation (Jensen 2001, 409). It is certain though, that a Neolithisation occurs among the existing population in Southern Scandinavia shortly after 4000 BC, although we must accept a certain influx of Neolithic people from the South.

Compared to Southern Scandinavia, the existing conditions on Shetland are rather different because Shetland does not have confirmed traces of a Mesolithic population which could turn Neolithic. In 2004-2005 though, very early dates of 4200-3600 cal BC were connected to a kitchen midden excavated at West Voe, Sumburgh (Melton 2009, 184; see also Melton and Nicholson 2007, 99; Melton personal comments), fig. 1. Even though none of the West Voe finds can tie the area to a Mesolithic environment, or a Neolithic one for that matter; the presence of hazel may indicate

the possibility of at least Mesolithic visits to the Shetland Islands (Edwards *et al.* 2009, 113; Edwards and Whittington 1998, 5). At the time of the Neolithic expansion c 4000 BC Shetland hardly contained a large Mesolithic population if any, so the Neolithic population probably expanded into empty islands in respect of a human population. As we shall see later that kind of expansion is not uncomplicated and certain conditions must be present for securing a successful expansion.

Why choose Shetland?

There are several reasons to choose Shetland as one part of a comparative analysis. First of all, Shetland is considered to be the northernmost area of Europe where farming was practised as a result of the expansion around 4000 BC. After a standstill of about 1000 years on the Continental lowlands just south of the Baltic, the farming economies expanded to the North and West. In Scandinavia, the expansion ebbed away at Svinesund on the border of present day Sweden and Norway 58,5o North, while it reached its ultimate boundary on the British Isles approximately 60o north on Shetland, and according to past maritime technology further expansion was impossible. Absolute dates from Shetland are infrequent, but the dates from Scord of Brouster, 3870 cal BC (Edwards and Whittington 1998, 14) and 4029 cal BC (Owen and Lowe 1999, appendix 7) have been confirmed as some of the oldest.

Shetland is moreover a closed unit, although it is scarcely also a closed laboratory. One of the tasks set by this research project will be to clarify possible outward relations after the islands were first colonised by the farming population. It is

Figure 1: West Voe, Sumburgh, Shetland. D. Mahler photo

evident that the Neolithic cultures on Shetland were related to the rest of the Neolithic traditions in North-West Europe, and it is just as evident that the Neolithic communities developed their own cultural traits with regards to tools, building traditions, and grave monuments.

Two cases of pioneer societies

This paper deals with peopling Shetland as there are different challenges to understanding the Neolithisation on the mainland on the British Isles and the continent and on islands especially of the size of Shetland and Orkney. I have been working with my project for 1½ years and have revealed that most of the elements mentioned in the introduction if not all are present in Shetland, which indicate that there was a well functioning society on Shetland during the Neolithic Period (Mahler 2011a; 2011b). This brings me to deal with the question of demographic aspects, wondering how island populations survive their initial colonization.

A pioneer society consisting of a 100 persons may well be economical successful, but if 98 of them are men and only 2 women the demographic prospective is prosperous. Besides population size we talk about three main elements, namely birth rate or fertility, death rate and sex ratio (Moore 2001a, 397; Moore and Moseley 2001b, 526; Robert-Lamblin

2006, 235). It is also important to remember that even fairly large population of let say 100 persons, males and females in the reproductive age will after some hundreds of years have great difficulties finding an acceptable marriage partner avoiding the society's definition of incest. These definitions could force the society to an exogamous marriage system (Moore 2001a, 406). Beside there is not a "magic number" for a population size that ensures a band's viability in a new environment.

"There is no such number. Initial size is only one factor contributing to the success of a colonizing group. Probably more important is birth rate, and good luck in having a balanced number of male and females born into the band..." (Moore and Moseley 2001, 528).

Generally we speak of six different models of colonization into empty areas, but here I shall only touch upon three of them (Moore 2001a, 395-396; Anderson and Gillan 2000, 43). The first model and much discussed is called Outpost Model or Leap-frog Model, which describe a human population colonizing an empty area without contact to the mother population. Such expansions are very vulnerable and constitute a demographic risk to say at least. This model is much more suited to situations where the expansion goes into already populated areas i.e. with an existing mass of genes, which could explain the rapid Neolithic expansion in continental Europe where the population of the Mesolithic of Ertebølle/ Ellerbech Cultures constitute possible marriage partners for the expanding TRB population (Rowley-Conwy ip, 1). The TRB expansion is a large and fast push only coming to a temporary halt just North of Bohuslän in Western Sweden as earlier mentioned, may be because the Mesolithic population density drops drastically further North.

The String of Pearls is a variation of the Matrix Model (Moore 2001a, 395) and describes a colonization along coastlines or river systems securing continuous communication with neighbouring sister settlements eventually securing this communication with exchange of e. g. raw material for making implements, as we shall see later with Sarqqaq Culture of Western Greenland (Grønnow 2004, 66). The last model to be mentioned is the Pulse Model or Wave Model, fig. 2. It is used for describing two beachhead and outpost scenarios, where colonists arrive in successive groups securing a steady gen flow. The successive arrives of new groups could be caused by special attractive elements in the newly colonized land such as arable land with the possibility of a social rise or special luxury raw material such as walrus teeth, if we consider Norse Greenland. In these far out societies it is necessary for some kind of social economic relations, which the archaeologist could recognise as exotic objects and the anthropologist e. g. ritualized objects. These phenomena could indicate a gene flow so to speak as the most important function. One of the most famous examples in the anthropological research is the Kula Exchange Ring and Red Feather Money both in Oceania (Malinowski 1922; Neich 2006, 217; Kirch 1988, 103). The Norse expansion in the North Atlantic during the 9th and 10th Century is an archaeological-historical example. Their dependence on an exchange system solving the demographic dimension for 500 years, resulted in demographic failure with total depopulation during the 15th Century for the Norse population in West Greenland (Mahler 2007, 412). The cause of this depopulation should probably be seen as part of demographic development in Scandinavia and Europe as a whole. After mid 14th Century there were waves of diseases not at least the Black Death, which reduced the European population with between 30 and 60 % (Lynnerup 1998, 122; 2011, 328). There were no pressure on the arable land resources any longer, and living on the fringe of the known world far away from family relations, it must have been much less attractive than gradually returning to where they had come some 500 years earlier. 500 years living on the fringe of the known world is also a kind of achievement, though.

The first paleo Eskimos expansion into the High Arctic started around 2500 BC, they crossed the straits between Ellesmere Island to the Northernmost Greenland, initiating the Independence 1 Culture, part of the Arctic Small Tool Tradition. The Independence 1 Culture expands to the East through Peary Land to areas with a rich population of Musk Oxen, but North East Greenland is at the same

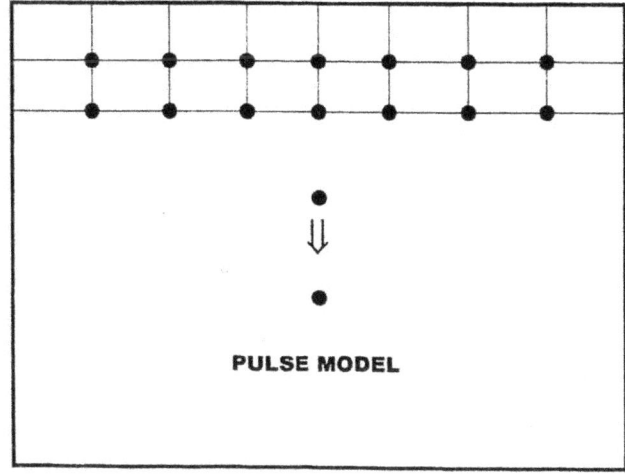

Figure 2: One of six demographic models. After Moore 2001, fig. 5.

time a cul-de-sac from where it is very difficult to return (Grønnow 2004, 78); by 1800 BC the expansion over and the depopulation a fact (Andreasen 2004, 62). The High Arctic is a severe challenge for any human population, and even small variations in climate and resources may be a threat. The demographic aspects could be an even worse challenge, as the small population maybe very thinly spread over huge distances, so without a steady influx of new genes the bands are doomed to extinction. In fact it is most impressive that the Independence 1 Culture in fact lasted between 200 and max. 700 years.

The Sarqqaq Culture, which is the archaeological name for a contemporary paleo Eskimo culture, did much better with a higher population density in the resource rich and varied Low Arctic area of West Greenland (Grønnow 2004, 66). The Sarqqaq Culture survived for over 2000 years, and the youngest traces of the culture are dated between 800 and 500 BC. The culture is characterized by bifacial knives, adzes, tanged points, micro blades and burins made of Killiaq, a grey metamorphosed slate (Sørensen and Pedersen 2005, 110; Grønnow and Sørensen 2006, 59), fig. 3. The stone quarries for killiaq is situated only in the Disko Bay area, but the killiaq raw material and artifacts made of killiaq are quite common on all the settlements of the Sarqqaq Culture often a very long way from the quarries. The exchange chains with killiaq could very well be a mediator for a gene flow along the West coast of Greenland, and thus creating a demographically successful development for the Sarqqaq Culture – for at least 2000 years.

The initial population in Western Oceania arrived some 30.000 year ago where they reached as far as the Solomon Islands but could not expand further because of the lack of maritime technology. It was impossible to expand out in the eastern Oceania bringing enough provisions at the same time by rowing alone. Only with the invention of sailing

Figure 3: Dark band of killiaq in the Østerfjeldet, Nuussuaq, Grønland. After Sørensen & Pedersen 2005:110.

technology did it become possible for further expansion around 2000-1500 BC (Irwin 2006, 62). The expansion of the Lapita Culture, a farming culture, could be due to a mixture of over population in the already colonized islands and the wish for social mobility. The Lapita expansion is a pulsating expansion with periods of stability changing with renewed pushes and periods of stand stills. The first push is supposed to have happened shortly after 2000 BC, reaching Hawaii around 700 AD and Easter Island 900 AD – an expansion covering one third surface of our planet (Anderson 2001,17; Kirch 2007, 332; Irwin 2006, 67). Some of the Islands became depopulated such as Henderson Island, Pitcairn, Necker and some of the small Polynesian islands, while other islands showed a considerable population growth such as Tikopia (Kirch 2007, 334; Firth 1959). Research into the exchange networks show connections between as many as 20 or 30 island societies through socio economic transactions. Kula and Red Feather Money have been mentioned, and though the systems are fairly recent maybe just two to three hundred years old, they show us the essentials of exchange with prestige objects as mediator for a gene flow. Archaeological excavations have shown much older systems among other places on Vanuatu, Vanikoro and Tikopia. Both Kirch and Friedman recognize the exchange importance of the systems for the social reproduction and the forming of marriages especially in the Eastern Lapita area (Kirch 1988, 114; Friedman 1981, 275).

Conclusion

We still need more research before understanding the Neolithic expansion at the ultimate European border around 4000 BC on Shetland. Is it an expansion which uses Orkney and Fair Isle as stepping stones? Or should we see the expansion as a Westerly phenomenon from the Hebrides? One thing is certain that the colonizing must have been pulsating that means it presumably consisted of waves of influx securing a demographic success, or the Shetland population had either connections back to the mother population or was part of an exchange system as we have seen above with Sarqqaq and in Oceania. This brings us in search for one or more mediators or special prestige items. One would expect that the beautiful, point butted felsite axes, totally polished and some of them never used or hafted could be items in an exchange chain and thus a mediator (Mahler 2011b, 61), fig. 4. On Shetland these long special axes are often found in wetlands, and seen from an archaeological point of view, they could be ritual deposits though we know precious little of the circumstances. But there are none of these axes outside Shetland – other than in museum collections. The same goes for the Shetland Knives, which are probably Late Neolithic, found packed together in numbers up to 19 and deposited in wetlands. As far as is known we have no objects with affinity to Shetland found in known Neolithic context on Orkney, Scotland or the Hebrides (Ballin 2011, 32). The Bronze

Figure 4: Long totally polished felsite axe from Shetland. There is no indication either of wear or of hafting. Courtesy Lerwick Museum.

Age on Shetland is very poor on actually bronzes, and finds from Jarlshof and the footmarks from Clickhimin suggest Westerly connections, maybe Irish (Turner 1998, 65). Whether these connections have older roots is not yet known, and the proof of a supposed network during the Neolithic period is still to be found.

Acknowledgements

Bjarne Grønnow has been of great help with discussions about the Greenlandic material and Poul Otto Nielsen, who is a lifelong researcher into the Neolithic, has helped in many other ways. Thanks to Martin Appelt for important references – all of them the National Museum of Denmark. Hans Christian Gulløv has provided important support since the beginning of Northern Worlds, The National Museum of Denmark. I would also like to thank my network in the Neolithic project Farming on the Edge. Shetland, the Border of Farming 4000-3000 BC Val Turner, Shetland Amenity Trust, Ian Tait and Jenny Murray both Lerwick Museum, Torben Ballin, Stirling, Flemming Kaul the National Museum of Denmark and Preben Rønne, Vetenskapsmuseet, University of Trondheim. And a special thanks to Susan Dall Mahler, who is a very patient person.

Bibliography

Anderson, A. 2001. Mobility models of Lapita mirgration. In Clark, G. R., Andersson, A. J. And Vunidilo, T.,(eds.) *The Archaeology of Lapita Dispersal in Oceania, Papers from the Fourth Lapita Conference, June 2000,*15-23. Canberra, Australia. Canberra. Terra Australis 17, Canberra.

Anderson, D. G. and Gillam, J. C., 2000. Paleoindian Colonization of the Americas: Implications from an Examination of Physiography, Demography and Artifact Distributions. *American Antiquity* 65, 43-66.

Andreasen, C., 2004. Independence 1-kulturen, In Gulløv, H. C. (ed.) *Grønlands forhistorie*, 204: 37-64. København, Gyldendal.

Ballin, T. B., 2011. The Post-Glacial Colonization of Shetland – Integration or Isolation. Evidence from Lithic and Stone Assembleges, Farming on the edge: Cultural Landscapes of the North. In Mahler, D. L. and Andersen, C., København (eds.) *Some features of the Neolithic of Shetland. Short papers from the network meeting in Lerwick, Shetland September 7th - 10th 2010. Northern Worlds.* 32-43. Copenhagen, The National Museum of Denmark.

Edwards, K.J., Schofield, J.E., Whittington, G. and Melton, N.D., 2009. Palynology "On the Edge" and the Archaeological Vindication of a Mesolithic Presence? The case of Shetland. In Finlay, N., McCartan, S., Milner, N. and Wickham-Jones, C. (eds.), *Bann Flakes to Bushmills, papers in honour of Professor Peter Woodman*, 113-123. Oxbow Books, Oxford.

Edwards, K.J. and Whittington, G. 1998. Landscape and environment in prehistoric West Mainland, Shetland. *Landscape History.* 20, 5-17.

Firth, R., 1959. *Social Change in Tikopia, Re-study of a Polynesian Community after a Generation*, London. London, London School of Economics.

Friedman, J., 1981. Notes on structure and history in Oceania, *Folk* 23, 275-295.

Grønnow, B., 2004. Saqqaqkulturen. Gulløv, H. C. (ed.), *Grønlands forhistorie*, 66-108. København, Gyldendal.

Grønnow, B. and Sørensen, M., 2006. Paleo-Eskimo Migrations into Greenland: The Canadian Connection, Dynamics of Northern Societies. Arneborg, J. and Grønnow, B., (eds.): *Proceedings of the SILA/NABO Conference on Arctic and North Atlantic Archaeology*, 59-74.

Grønnow, B., In press. Independence I and Saqqaq: the first Greenlanders. Friesen, M. and Mason, O. (eds.) *Handbook of Arctic Archaeology*. 1-23, Oxford University Press, Oxford, New York.

Jensen, J., 2001. *Danmarks Oldtid*. København. Gyldendal.

Irwin, G., 2006. Voyaging and Settlement, Vaka Moana, In Howe, K. R., (ed.). *Voyages of the Ancestors, the Discovery and Settlement of the Pacific*, 56-99. David Bateman, Auckland Museum, Auckland.

Kirch, P. V., 1988. Long-distance exchange and island colonization. The Lapita case. *Norwegian Archaeological Review* 21(2), 103-117.

Kirch, P. V., 2000. *On the Road of the Winds: An Archaeological History of the Pacific Island Before European Contact*. Berkeley. University of California Press.

Kirch, P. V., 2007. Concluding Remarks. In Kirch, P. V. and Rallu, J-L. (eds.), *2007: The Growth and Collapse of Pacific Island Societies*, 326 – 337. Honolulu. University of Hawai'i Press

Lynnerup, N., 1998. The Greenland Norse. A biological-anthropological study. *Meddelelser om Grønland: Man and Society* 24, 1-149. Copenhagen, The Commission for Scientific Research in Greenland.

Lynnerup, N., 2011. When population decline. End period demographic and economics of the Greenland Norse. In Meier, T. and Tillessen, P. Hrsg. (eds.), Über *die Grenzen und zwischen den Disziplinen. Fächerübergreifende Zusammenarbeit im Forschungsfeld historischer Mensch-Umwelt-Beziehungen.* Archaeolingua Alapitvany, Budapest: 335-345.

Mahler, D. L., 2007. Sæteren ved Argisbrekka. Økonomiske forandringer på Færøerne i vikingetid og tidlig middelalder. Economic development during the Viking Age and Early Middle Ages on the Faeroe Islands. *Annales Societatis Scientiarum Færoensis Supplementum*, Tórshavn, Faroe University Press: 1-525.

Mahler, D. L., 2011a, Shetland: The Border of Farming 4000-3000 BC. Features of the Neolithic Period on Shetland, In Mahler, D. L. and Andersen, C., København (eds.), *Farming on the edge: Cultural Landscapes of the North. Short papers from the network meeting in Lerwick, Shetland September 7th - 10th 2010.* Northern Worlds, 6-18.

Mahler, D. L., 2011b, Shetlandsøerne: Landbrug på grænsen 4000-3000 f.v.t. Nogle træk af yngre stenalder på Shetlandsøerne. Ændringer og udfordringer, In Gulløv, H. C., Paulsen, C. and Rønne B. (eds.), *rapport fra Workshop 1 på Nationalmuseet, 29. september 2010, Nordlige Verdener*, København, 59-68.

Malinowski, B., 1922: *Argonauts of the Western Pacific*. London. Percy Lund, Humphries & Co, London.

Melton, N.D. 2009. Shells, seals and ceramics: an evaluation of a midden at West Voe, Sumburgh, Shetland, 2004-5. In MaCartan, S., Schulting, R., Warren, G. and Woodman, P. (eds.), *Mesolithic Horizons, Papers presented at the Seventh International Conference on the Mesolithic in Europe, Belfast 2005.* 1, 184-189.

Melton, D.N. and Hicholson, E.A., 2007. A Late Mesolithic – Early Neolithic midden at West Voe, Shetland. In Milner, N., Craig, O. E. and Baily, G. N. (eds.), *Shell Middens in Atlantic Europe*. Oxbow Books, Oxford: 94-100.

Moore, J. H., 2001. Evaluating Five Models of Human Colonization. *American Anthropologist* 103(2), 395-408.

Moore J. H. and Moseley M. E. 2001. How many Frogs does is take to Leap around the Americas? Comments on Anderson and Gillam, *American Antiquity* 66, 526-529.

Neich, R. 2006: Voyaging after the Exploration Period, Vaka Moana, Voyages of the Ancestors. In Howe, K. R. (ed.), *The Discovery and Settlement of the pacific*, 198-245. David Bateman, Auckland Museum, Auckland.

Owen, O. and Lowe, C., 1999. Kebister: The Four-thousand-year-old Story of One Shetland Township. *Society of Antiquaries of Scotland Monograph Series, Number 14.* Edinburgh. Historic Scotland.

Robert-Lamblin, J., 2006. Demografic Fluctuations and Settlement Patterns of the East Greenlandic Population – as Gathered from Early Administrative and Etnographic Sources, Dynamics of Northern Societies, In Arneborg, J. & Grønnow, B., (eds), *Proceedings of the SILA/NABO Conference on Arctic and North Atlantic Archaeology.* 235-246. Copenhagen.

Rowley-Conwy, P. In press. Westward Ho! The Spread of agriculture from Central Europe to the Atlantic, *Current Anthropology*, 1-29.

Sørensen, M. and Pedersen, K. B., 2005. Killiaq kilden på Nuussuaq: på sporet af Saqqaqkulturenredskabssten. *Tidsskriftet Grønland* 2-3, 105-118.

Turner, V. 1998. *Ancient Shetland*. Historic Scotland. Shetland Amenity Trust. D. J. Breeze ed. Historic Scotland, Shetland Amenity Trust, B. T. Batsford Ltd. London: 1-127.

CHAPTER 7
BRINGING IT ALL TOGETHER – PITS AS MONUMENTS

Anna-Karin Andersson

Linnaeus University

Abstract: *The article sets out to explore the relationship between Mesolithic and early Neolithic coastal sites, such as early Neolithic inland sites and long mound sites in the southern part of Scania, in Sweden. Focus is placed on pits and depositions at these sites. The author argues that the very construction of pits and the variety of depositions within them should be seen as bringing practical action into a symbolic room. In that sense, pits should not be divided into sacrificial deposition and non-sacrificial deposition. Instead, we should simply notice that putting things into the ground and making a closure, has contributed effectively to the rise of monumentality. In Scania, pits are, in several cases, present at sites with long mounds and therefore, needs to be understood as incorporated into the very process of monumental development.*

Keywords: *monuments, pits, Neolithic, Mesolithic, coastal sites, deposition*

The argument that pits needs to be incorporated into the very process of monumental development is based on a three-fold idea deriving from my thesis; 1) coastal so called mixed sites with material from the Ertebølle culture (EBK) and the Funnelbeaker culture (TRB) (Jennbert 1984; 1985; Petersson 1999; Jonsson 2006; Andersson 2011 forthcoming), 2) early Neolithic inland sites (Larsson 1984; Rudebeck and Nilsson 2010) and 3) sites with long mounds and early Neolithic pits (Rudebeck and Ödman 2000; Rudebeck and Nilsson 2010; Gidlöf 2006; 2009).

The coastal sites displays complex stratigraphic sequences derived from the period around 4000 BC. Depositional signature is of regular use, most likely seasonal and cumulative, representing the passing of time in seasonal cycles. Cultural material is being dumped and cast-off. This is a 'casual' discard of things in the context of sand accumulation resulting from some wind-blow and fluvial deposition. The stratigraphy has been subjected to criticism with critics claiming a mixing of strata due to fluvial actions.

In some cases, this mixing is demonstrably false; e.g. Vik on the eastern coast of Scania, that is situated 17 meters above sea level and could not have been affected by the ocean waves (Strömberg 1965; 1973; 1986; Helgesson and Björk 1998; Andersson 2011 forthcoming). One of the more famous sites is Löddesborg (Jennbert 1984; 1985) on the western coast of Scania that Kristina Jennbert wrote her dissertation on in the mid 1980s. Another site is situated not too far from Löddesborg, named Järavallen and was excavated in the beginning of the 20th century (Kjellmark 1907), right next to this one is Elinelund, a site of similar character (Sarsnäs and Nord Paulsson 2001; Jonsson 2005).

At Elinelund, a stone paved pit with a piece of an axe displaying several feature of a shoe-last adze was present (Jonsson 2005). The oldest part of Elinelund is situated right next to the former coast and the stratigraphy was consistently deposited through time, which makes it possible to follow the cumulative process from the Mesolithic to the Neolithic. There is no evidence for substantial abandonment of the site (cf. also Jennbert 1984 regarding Löddesborg for a similar discussion). A coarse type of pottery made in a very rough N technique, a kind of a transitional technique between the H and U techniques mainly associated with the Mesolithic and the N technique in the Neolithic (Rudebeck and Ödman 2000; Jonsson 2005), was found at the site and is similar to the kind of pottery present at Löddesborg (Jennbert 1984). This kind of transitional pottery is highly informative in the debate concerning how old traditions have gradually been replaced by newer ones. Similar pottery has been found at a few inland and long mound sites as well.

In that way and regardless of the integrity of the stratigraphy, sites like these are of importance because they demonstrate how a slow, gradual change comes into being.

The Neolithic Inland Sites

It is during the same period, or slightly later, that the mixed sites on the coast are in use; this corresponds with a tendency of increasing settlements inland (e.g. Larsson 1984). These, are often very small sites differing from the coastal sites in that they are exclusively early Neolithic in character. At several of these sites, characteristic early Neolithic pits are present (cf. for example Ericsson *et al.* 2000; Rudebeck 2010) filled with material remains, everything from Neolithic pottery and grains to flint and stone tools.

One example of an inland site is Kabusa II, a site with two recorded pits in its outskirts that contained varied material with pottery and flint. Another example is from Käglingevägen. The site is located in a hilly landscape and nearby a former wetland. Here, a cultural layer was present along with some pits. A vessel of the earliest Neolithic character was reconstructed after being found in a pit at Kabusa II and flint artefacts and debris were found both underneath and on top of the pottery (see Larsson 1984).

My research has revealed the special role that pits and the material deposited within them had in defining the character of coastal sites, the Neolithic inland sites and the site for the first monuments in Scania - the Neolithic long mound (see for example Rudebeck and Ödman 2000; Gidlöf *et al.* 2006; 2009; Rudebeck 2010). Authors have attempted to distinguish between votive or ritual deposits in pits, and the more mundane dumping of debris (Bradley 1998; Chapman 2000; Chapman and Gaydarska 2007; Thomas 1999; Jones 2007; Larsson 1997 etc.). But I would like to put these distinctions aside (all activities are likely to be imbued with various values and meanings that now lie beyond our recovery), and notice simply that to deposit material in the ground, rather than as a midden on the surface as we see later on with the long barrows, marks the landscape in certain ways. Instead the ground is broken open, this becomes a clearly defined focal point for activity, and material is gathered from a number of sources and places, and hidden in the permanence of the earth's surface (Andersson 2011). Memory is sealed and secured. The whole act of putting things into the ground is closure of the life-cycle of the artefact. This final act equally finishes the connection and interaction of the people that have been involved in the life of the deposited artefact. What depositions in pits do is marking the landscape in specific ways; choosing to end the life of artefacts at a specific location connects back to the artefact's past, the past of a place and the past of the society/groups.

Early Neolithic pits, often with a rich inventory have been interpreted for numerous purposes; rubbish pits, storing pits or depositional pits (Larsson 1984; 1997; Erikson *et al.* 2000; Gidlöf *et al.* 2006; Rudebeck 2010). At Almhov in eastern Scania, over 130 pits were found during the excavation. Typologically and radiocarbon places these pits in the very early Neolithic (Gidlöf *et al.* 2006:2009). Of the 130 pits, 94 were categorized as "early Neolithic rich-in-find-pits" (see Eriksson *et al.* 2000). Some of the pits contained up to 70 kg of material including flint, bone and pottery. A couple of twin pits were found that were aligned in pairs, often next to each other containing broken parts of the same object. Gidlöf *et al.* (2006; 2009, 53) separates between ritual or sacrifice pits and other pits. He also identifies three pits which stands out as pits for sacrificing but acknowledge the fact that it is very hard to determine which one of the pits should be considered ritual and which one that should not. The pits that are considered for sacrificial use are those that contain artefacts that really stands out from the others, such as pit A866. This pit consists of several depressions containing pottery, flint, stone and animal bones. The flint-material amounted to 33 kg and the pottery was composed of at least 13 different kinds of vessels and four clay-disks (for further discussions

about the pits and the site of Almhov, see below). The idea of interpreting some pits as ritual and not others (Gidlöf *et al.* 2006:53; Eriksson *et al.* 2000; Garrow 2006) raises at least two important issues: Practical objects and habitual actions posses in themselves a symbolic value (e.g. Dobres 2007). The action of digging a pit is not just habitual but marks the site and the landscape in new ways. The very deposition of objects can be seen as mediators between the past, the present and the future of a community.

Therefore, the very construction of pits and the varying deposition within them could be seen as bringing practical action into a symbolically important room. There is no reason to treat just some of the pits as sacrificial depositions; even though those pits are the only ones where there is a clear tendency that the artefacts have been specially selected.

> *"It can be maintained that pit-digging, especially on a previously occupied site, constitutes an exchange between the living and the ancestors. In this sense the deposition of figurines in pits connects the living and the ancestors through figurine fragments, matching fragments of which are enchained to those same ancestors in whatever other context they may be used or deposited."* (Chapman 2000, 72)

Neolithic pits have also been recognized as a common feature outside Scandinavia, for example in Great Britain (e.g. Thomas 1999; Garrow 2006; Lamdin-Whymark 2008); where they are interpreted as being everything from waste pits to depose of artefacts to ritual pits. In Scandinavia, they have been debated for quite some time (e.g. Larsson 1984; 1997; Ericsson *et al.* 2000; Gidlöf et al. 2006; Nilsson and Rudebeck 2010). However, no further conclusions other than the same interpretations as above have been made - pits have been dug for various reasons spanning the field from waste to ritual deposits.

Without continuing an endless discussion of what each pit can be interpreted as, we can say that pits lie at the beginning of recognition that the accretion of human actions changes the form of the world (e.g. Barrett 2006). This did effectively contribute to the beginning of monumentality.

Pits and monumentality

All of the sites with long mounds in Scania, except for the Giant Grave outside Trelleborg, display evidences for pits placed in close vicinity to the long mounds. Several of the Scanian sites contain deposits that pre-date the long mound and an increasing frequency of deposits occurring in proximity to the monuments is observable. In that sense it is interesting to look at what Duncan Garrow (2006, 34) writes "...the relationships between pits and other features (whether monuments, buildings or scatters of post-holes) is difficult to assess." Garrow continues the argument with the observation that pits ought to be understood as archaeological features in their own right and need not to be interpreted along with monuments or other buildings. It is a relevant observation that pits could be features in their own right, but in western Scania, pits do serve as foundations for the first monuments; they precedes the mounds and the mounds are aligned accordingly to the pits.

As contradictory to Garrows (2006) arguments; early Neolithic pits cannot be interpreted as features in their own right, but must be understood as components in monumental buildings. In which the deposits, the backward-looking past forms the foundation for the forward-looking monument building.

With that said; what we are seeing in western Scania during the first centuries after 4000 BC are the indications of what communities were building for future conditions. Neolithic pits should not be considered as monuments though (in the meaning of the word that a site or a place is marked and preserved), but monumental (in the meaning of the word as "of outstanding significance"). If pits were marked above ground, with standing posts, façade like structures, monoliths or standing stones they ought to have been considered as monuments within their own right. Putting up a marker means that they are meant to be preserved and to be seen by others not just the own community and also to preserve for the future. Whilst depositing things in the ground without markings projects backwards. In other words, pits holding a deposit equate to a closure, in a sense the end of its journey, while pits that hold a post or monuments serves as a foundation for the future. Since pits sometimes occur along side long mounds, they ought to be understood as being incorporated into the very process of monumental development. This is demonstrated at a couple of sites in Scania.

Kristineberg

The site was excavated during the 1970s. It held the remains of two ploughed out long barrows and a number of pits. Two of the pits were features 84 and 87. Both could typologically, from the inventory, be placed in the early Neolithic. No 84, measured 1.6 by 1.1 metres, filled with sooty soil that held 238 grams of worked flint and waste from manufacturing. One flake, 10 cm long, with retouch on both sides was found along with a fragment of a polished axe and a fossilized sea urchin. The pottery in the pit derives from at least five different pots and is associated with the very early Neolithic Oxie and Svenstorp group (Rudebeck and Ödman 2000).

Pit 167, just north of one of the two long barrows was filled with one coherent layer of sooty sand. Pottery sherds of very early funnel beakers and a sherd of a very coarse tempered ware, 13 mm thick made in rough N technique similar to the pottery from Elinelund (mentioned above) were found. The pit was radiocarbon dated from charcoal to 4327 – 3985 BC (sigma 1 LuA-4538) which is one of the earliest dates for a funnel beaker assemblage in Scandinavia and comparable to other sites in Scania, for example Mossby

(Larsson 1997). Even more interesting was the impressions of emmer present in the pottery (Rudebeck and Ödman 2000). At this site pits are contemporary or slightly older than the actual long mounds and are to be interpreted as monumental features in the ground bringing artefact and material to a closure.

Almhov

The site of Almhov was excavated as part of the infrastructure work for the City Tunnel in Malmö in southeast Sweden. The site is located some kilometres from present day shoreline and is partly situated on a shelf that slopes down to the former wetland of the Hyllie bog. The actual site comprised at least four possibly five long mounds, and hundreds of pits, out of which over a hundred can be dated to the earliest part of the Neolithic. A total of 94 pits with very rich contents was dated to the early Neolithic I (Gidlöf *et al.* 2006; 2009; Rudebeck and Nilsson 2010). The pits contained pottery sherds, flint, stone artefacts and bones of animals. Due to the rich presence of pits the site has been interpreted as a site that functioned as a place for feasting and communal gatherings even before the long barrows were constructed (Gidlöf *et al.* 2006; 2009; Rudebeck and Nilsson 2010). The following sequence of the site is therefore:

1. A gathering/feasting site (when the construction of the pits occurred).
2. A more monumental and ritualized site (when the long barrows were constructed).
3. Broken discontinuity somewhere at the end of the early Neolithic and at the beginning of the middle Neolithic.

The pits were dated mostly with typology as a lot of the vessels belonged to the Oxie group, placing the pottery in the early Neolithic. This was also confirmed with radiocarbon dating (e.g. Lab.no. Ua-23871 charcoal of ash tree from pit A14685 which gave cal 1; 3970 – 3800 BC and the pit next to this one interpreted as a twin-pit to A14685, A14736 which gave cal. 1; 3940-3710 BC (lab. No. Ua-23872)).

It has been suggested that a decline in the frequency of rich pit finds seems to be the case when monuments begins to be built (e.g. Eriksson *et.al* 2000; Andersson 2011). That the monument itself takes over the role. This is observable at Almhov where the main sequence of pit assemblages can be associated with activity that pre-dates the long mounds. In some respects it seems as that the building of the long mounds have replaced the performances of depositing artefacts in pits or in other words that the future takes over from the past.

One example of early Neolithic pits at Almhov is feature 19049 located c. 30 meters south of the large long mound and just a few meters from the façade of another of the four long mounds at the site. Pit 19049 measured 4.9 by 3.9 meters and had a depth of 0.74 meters. This makes it the biggest at the site and as it contained pottery from at least 60 vessels along with five clay discs it was also one of the richest. The flint material alone exceeded 20 kg and most of it was waste and debris from artefact production.

The pit was stratified into 12 layers, suggesting that it had been held open for a while, how long of course is impossible to say. The pit had sloping edges and the bottom displayed an irregular profile that seemed to follow a source of clay in the ground. This caused the excavators to interpret the pit as originally being dug for clay collection perhaps for sealing the walls of houses (Gidlöf *et. al* 2006). In the north of the pit there was a stone packed posthole that had held a large post.

Examination of the other artefact categories present in the pit suggests the pit was use in the very first centuries after 4000 B.C. The long mound no.1, was radiocarbon dated from grain found in one of the post hole of the façade to cal. 3940-3660 B.C. (Ua-17158) and the pit was radiocarbon dated by samples taken from wheat found within layer 11 (quite far up) cal. 3960-3790 B.C. (Ua21383). That would make the two structures more or less contemporary.

The question is not whether the interpretation of the pit's original - being dug for a clay source - is correct or not, the focus ought instead to be placed on the fact that there is no contradiction between taking up clay and depositing household material in pits; there is simply a life cycle, making living (manufacturing pots and other tools) and dying (bringing material to a closure).

Discussion and conclusion

At the early Neolithic/long mound sites two fundamentally different processes are taking place, which involve the human recognition of time. First are pits, which mark the end of life and artefact circulation, in a sense the end of the artefact's biography. Second are monuments, which are constructed for the future but reference and point back to the past. Therefore the inland occupants place themselves in the cultural construction of a complex temporal landscape.

Pits do not contain ritually articulated assemblages, as some would like us to believe, but the end of a routine. The material in pits is in most cases, rubbish gathered together to look back on what life has been and to bury it. The various reasons that people deposited artefacts, used or unused, within pits is likely to fall outside our ability to investigate. To be able to better understand pits and depositions, there is a need to investigate the very action of digging pits and to start to recognize them as monumental features within their own right. Monuments above ground incorporate undisputable diverse elements from surrounding landscapes as well as from various aspects of individual and communal lives. But, pits do that as well; the difference lay in that pits projects backwards and standing monuments projects forward. In other words; pits create a closure and standing monuments a future. But for a future a closure is needed.

The coastal sites, the early Neolithic inland sites and sites with long mounds are linked together but also do three profoundly different things. The coastal sites involve human as well as natural agency - people continuing with their daily life. In this context we see the steady transition from EBK to TRB but crucially not the rapid displacement of the earlier by the later material. There is instead an accumulation marking out a very complex process of cultural evolution during which cultivation has developed.

The Neolithic inland sites and sites with long mounds bring together the history of the landscape and the history of the people in depositions, pits and monuments. The human agency reveals in this way an underlying order of the world and acknowledges how that order becomes explicit in the formation of landscape. Therefore, the digging of pits and the building of long mounds in Scania should be understood as actions that bring forward a past, highlights the present and take it all into the future.

References

Andersson, A-K. 2011. Tracing the future in the past – tracking change in a local perspective. In I. Hadjikoumis, A. Robinson and S. E. Vyner (eds.), *The dynamics of Neolithisation in Europe: Studies in honour of Andrew Sherratt*, 353-363. Oxford, Oxbow.

Barrett, J. 2006. A Perspective on the Early Architecture of Western Europe. In J. Maran, C. Juwig, H. Schwengel and U. Thaler (eds.), *Konstruktion der Macht: Architektur, Ideologie und soziales Handlen*, 15-30. Hamburg, Bradley.

Bradley, R. 1998. *The Significance of Monuments On the shaping of human experience in Neolithic and Bronze Age Europe*. London, Taylor and Francis Ltd.

Bender, B. 1978. Gather-hunter to farmer: a social perspective. *World Archaeology* 10(2), 204-222.

Chapman, J. 2000. Fragmentation in archaeology. People, places and broken objects in the prehistory of South Eastern Europe. London, Routledge.

Chapman, J and Gaydarska, B. 2007. *Parts and wholes. Fragmentation in prehistoric context*. Oxbow, London.

Dobres, M. A. 2000. *Technology and social agency: outlining a practice framework for archaeology*. Wiley-Blackwell, Oxford.

Edmonds, M. 1999. *Ancestral Geographies of the Neolithic: landscapes, monuments and memory*. Routledge, London.

Eriksson et al. 2000. Fyndrika TN-gropar i sydvästra Skåne. CD-uppsats i arkeologi, Lunds universitet, Lund. Unpublished MA Thesis, Lund University.

Faris, J. C. 1975. Social evolution, population and production. In H. Lumley (ed.), *Population, Ecology, and Social Evolution*, 235-277. De Gruyter Monton, Hague.

Garrow, D. 2006. *Pits, Settlements and Deposition during the Neolithic and Early Bronze Age in East Anglia*. BAR series 414, Oxford.

Gidlöf, K., Hammarstrand, K. Dehman K., Johansson, T. 2006. *Almhov – delområde 1. Malmö: Malmö kulturmiljö* Rapport 39 2006. Malmö Kulturmiljö, Malmö.

Gidlöf, K. 2009. En tidigneolitisk samlingsplats – fyndrika gropar och långhögar på Almhov. In M. Hadevik and C. Steineke (eds.), *Tematisk rapportering av Citytunnelprojektet: rapport* över *arkeologisk slutundersökning*, 91-136. Malmö, Museer Malmö.

Helgesson and Björk 1998: B. Helgesson / T. Björk, Rapport. Skåne, Kristianstads län, Simrishamns kommun, Simrishamn, Kv Lars-Johan 4, 15 och 16. Arkeologisk för och slutundersökning 1987. Fornlämning 25. Opubl. Rapport. Länsmuseet i Kristianstad. Kristianstad.

Ingold, T. 2000. *The Perception of the Environment. Essays in livelihood, dwelling and skill* Routledge, London.

Jennbert, K. 1984. *Den produktiva gåvan. Tradition och innovation i Sydskandinavien för omkring 5300 år sedan.* Acta Archaeologica Lundensia. Series in 4°, Lund.

Jones, A. 2007. *Memory and Material Culture.* Cambridge, Cambridge University Press.

Jonsson, E. 2005. Öresundsförbindelsen. *Skjutbanorna 1A : rapport* över *arkeologisk slutundersökning.* Malmö, Malmö Kulturmiljö.

Kjellmark, K. 1903 Antiqvarisk tidskrift för Sverige. *En stenålderboplats vid Järavallen vid Limhamn*, 17 (3), 1903.

Lamdin-Whymark, H. 2008. *The Residue of Ritualised Action: Neolithic Deposition Practices in the Middle Thames Valley.* BAR series 466, Oxford.

Larsson, M. 1984. *Tidigneolitikum i Sydvästskåne. Kronologi och bosättningsmönster.* Acta Archaeologica Lundensia Se- nes in 4°. N° 16, Lund.

Larsson, M. 1990. Det sydskånska backlandskapet – stenåldersjägaren blir bonde. Genetik och Humaniora 2. *Meddelanden från Erik Philip-Sörensens Stiftelse för främjandet av genetisk och humanistisk vetenskaplig forskning.* Lund.

Larsson, M., Olsson, E., Biwall, A. 1997. *Regionalt och Interregionalt. Stenåldersundersökningar I Syd- och MellanSverige.* Riksantikvarieämbetet, Stockholm.

Lindahl, A., Olausson, D., Carlie, A., Stilborg, O. 2002. *Keramik I Sydsverige: En handbok för arkeologer.* Lund, Keramiska forskningslaboratoriet.

Papmehl-Dufay, L. 2006. *Shaping an identity; Pitted-Ware pottery and potters in south-east Sweden.* Stockholm, Stockholms Universitet.

Petterson, H. 1999. *Nationalstaten och arkeologin – hundra år av neolitisk forskningshistoria och dess relationer till samhällspolitiska förändringar.* Göteborg, Institutionen för arkeologi Göteborgs universitetet.

Price, T. D. and Brown, J. A. 1985. *Prehistoric hunter-gatherers: the emergence of cultural complexity.* Academic Press, Orlando.

Rudebeck, E. 2010. I trästodernas skugga – monumentala möten i neolitiseringens tid. In E. Rudebeck and B. Nilsson (eds.), *Arkeologiska och förhistoriska världar : fält, erfarenheter och stenåldersplatser i sydvästra Skåne*, 83-251. Malmö museer, Malmö.

Rudebeck, E. and Ödman, C. 2000. *Kristineberg – en gravplats under 4500 år. Stadsantikvariska avdelningen kultur.* Stadsantikvariska avdelningen, Malmö.

Sarsnäs, P. and Nord Paulsson, J. 2000. Öresundsförbindelsen. *Skjutbanorna 1B and Elinelund 2A. B 8* rapport över arkeologisk slutundersökning. Malmö, Malmö museer.

Strömberg, M. 1965. *An Early Neolithic Settlement Site. Meddelanden från Lunds Universitets historiska museum 1964-1965.* Lund, Gleerup.

Strömberg, M. 1973. Rapport för kv. Lars-Johan 5, Dnr 8142 173. Simrishamn, Skåne, ATA.

Strömberg, M. 1986. *Signs of a Mesolithic Occupation in South-East Scania. Meddelanden från Lunds universitets historiska museum 1985-1986. vol 6.* Stockholm, Almqvist and Wiksell.

Thomas, J. 1999. *Understanding the Neolithic.* London, Routledge.

Chapter 8

Disclosing the World During the Mesolithic/Neolithic Transition in the Irish Sea Basin

Hannah Cobb

University of Manchester

Abstract: *Interpreting the transition from hunting and gathering to farming in the UK is highly problematic. Contrasting theoretical traditions have predominated in Mesolithic and Neolithic studies in the present providing a disjointed view of the transition between the two in the past. To move beyond these contrasts this paper examines how Heidegger's notion of disclosure may be of value for interpreting the transition from the Mesolithic to the Neolithic period. Such an approach considers how people, places and things show up and how they gather the world around them. During such a crucial time of transition, exploring such connections enables us to move beyond traditional functional and economic studies that predominate in Mesolithic studies, and instead explore the complicated intersections of place, materials, material practices and the production of identity that took place at the Mesolithic/Neolithic transition..*

Keywords: *Mesolithic, Neolithic, Transition, UK, Phenomenology, Disclosure, Materials/Materiality, Place, Identity*

Introduction: Problems with interpreting the Mesolithic/Neolithic transition in the Irish Sea basin

In this paper I will consider the Mesolithic/Neolithic transition in the Irish Sea Basin (Figure 1), an area that has been the subject of my doctoral and post-doctoral research since 2004. The Irish Sea basin has been the focus for a range of Mesolithic and Neolithic studies in recent years (Cobb 2008; papers in Cummings and Fowler 2004; Cummings 2009; Schulting 2004; Sheridan 2004, 2007, 2010) and as a result it has also been the focus of recent vigorous debate on the transition from hunting and gathering to farming. This debate revolves around two relatively polarised stances, one which views the transition as driven by incoming Neolithic populations from the continent (Rowley-Conwy 2004; Sheridan 2003; 2004; 2007; 2010), and one which views the transition as a largely indigenous change (Cobb 2008; Thomas 2003; 2004a; 2007; 2008).

However in this paper I argue that before we can resolve the question of how the transition occurred, there are a series of more fundamental issues that need to be addressed. In particular the three problems that I identify are:

The predominance of socially impoverished narratives of the Mesolithic period

This arises as a result of the predominantly processual, environmental and functional approaches towards the Mesolithic in the area. Of course a number of new, interpretive approaches to the Mesolithic in Britain are now being developed (e.g. papers in Bevan and Moore 2003; papers in Cobb et al. 2005; Cobb 2007; 2008; 2009a; 2009b; 2012; papers in Conneller 2000; Conneller 2004;

2011; papers in Conneller and Warren 2006; papers in Young 2000). Yet, in contrast to Neolithic studies (which in the UK are almost entirely undertaken under the interpretive or post-processual agenda), the Mesolithic continues to appear as a chronological period where people only had social relationships with hazelnuts (Bradley 1984) in many narratives, whether or not it really was. Moreover, the types of sites that have traditionally lent themselves to more socially interpretive accounts in Neolithic studies, such as chambered tombs, are not found in the Mesolithic record, with the exception of a small number of exceptional shell midden sites such as those on Oronsay in Western Scotland. This lack of overt "ritual" and symbolic material culture makes any approach to the Mesolithic from an interpretive stance more challenging and in turn further compounds the lack of socially situated narratives of the Mesolithic. Finally, whilst Neolithic material culture and site types are relatively homogenous within the area, this is much less the case in the Mesolithic. As a result, whilst broader statements can be made about Neolithic belief systems, once again for Mesolithic studies such interpretations remain challenging and often site specific, and as a result this only contributes to the lack of socially situated narratives of the period. As a result the different narratives in the present surrounding the two periods create a sense of significant difference between the social lives of Mesolithic and Neolithic people. Yet this narrative is almost entirely a product of modern paradigmatic differences, rather than a reflection of actual differences in the past.

The notion of an insular Mesolithic in the area

There is a notable lack of homogeneity in the Mesolithic material culture of the Irish Sea basin which is reflected in a number of ways; predominantly, differing lithic traditions

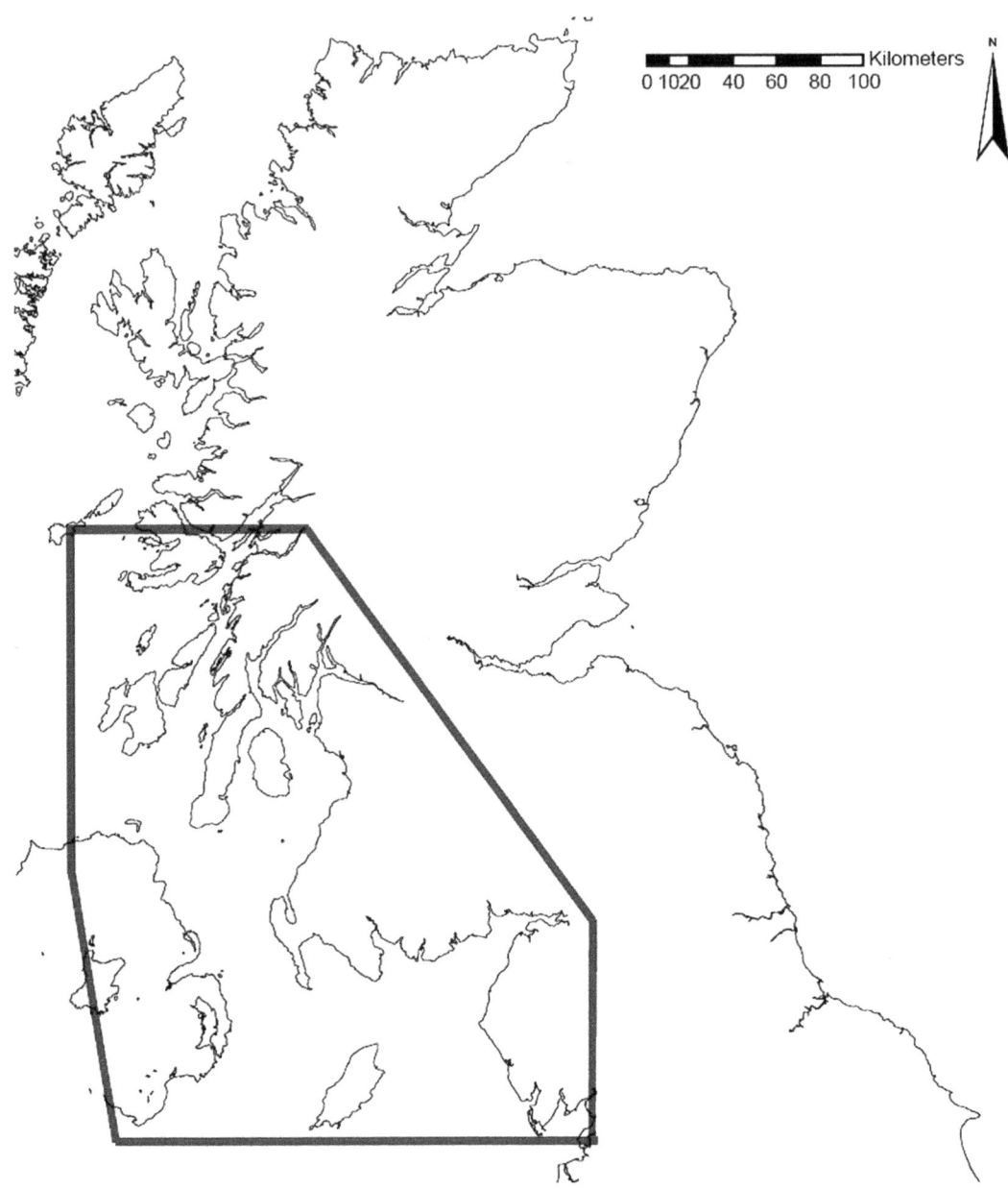

Figure 1: The area of the northern Irish Sea basin discussed in this paper (Source: Cobb 2008)

in Ireland and the Isle of Man, compared to those in Scotland and northern England, and a lack of movement of raw materials across the area have all been drawn upon as evidence for an insular Mesolithic (Woodman 2004). This in turn significantly contrasts with the Neolithic period where the movement of raw materials, tool types and the homogeneity in sites such as tombs all adds further to the seeming contrast between the two periods. However I have argued elsewhere (Cobb 2007; 2008) that once again this is symptomatic of differences in modern approaches. Whilst the Neolithic is often examined across the Irish Sea Basin (e.g. Cummings 2009), Mesolithic studies are more often regionally focussed – a tradition that dates back to the antiquarian studies of the late nineteenth century and the impact of geopolitics in the twentieth century (Cobb 2008). As a result scholars very rarely examine the Mesolithic of

the Irish Sea Basin as a whole in the way that they do for the Neolithic, and so once again the Mesolithic appears to have been a time of very different, insular lifeways in contrast to the Neolithic. Yet at least part of this narrative of Mesolithic insularity is a product of modern traditions of study. Moreover, as it has been argued elsewhere (Cobb 2008; 2009b; Thomas 2004), and demonstrated in ethnographic studies (Hodder 1982) a lack of homogeneity in material culture does simply equate to insularity and a lack of contact across the area.

The lack of sequences between the Mesolithic and Neolithic in the area

The final and very serious problem in the area is that there are very few transitional sequences in the archaeological record. However where we do get some semblance of transition is at exceptional sites such as shell middens. This in itself poses a problem as it has meant that archaeologists have drawn upon these vastly exceptional and extraordinary sites to inform our understanding of the transition, and this can in turn translate to a simple narrative in which shell middens are simply seen as precursors to megaliths (Cummings 2001; 2003; 2007; Pollard T. 1996; Pollard J. 2000). Yet the process of transition is one that is much more likely to have taken place and affected people through the routines of daily life.

The three issues outlined here all demonstrate that a serious problem exists in how the Mesolithic in the area is understood – i.e. often as insular and socially deprived. This then forces an interpretation of great contrast between Mesolithic and Neolithic lifeways which to a significant extent is a product of differences in modern theoretical paradigms. This in turn then affects how the transition in the area can be interpreted, and this is compounded by the use of predominantly exceptional sites to interpret the transition, rather than the sites that people lived much more of their lives through.

In reality, after collating details of all the Mesolithic site types in the northern Irish Sea basin (Figure 1) my doctoral research demonstrated that less than 3% of these sites were middens. The majority were instead lithic scatters – that is to say the majority of sites were the kind of sites that reflected daily life rather than exceptional and unusual acts of deposition of marine materials. Moreover 23% of Mesolithic sites in the study area either had a definite or probable Neolithic presence, and more specifically 9% of sites definitely had a Neolithic presence. In most cases this was represented simply as diagnostically Neolithic lithic typology or pottery, but in some cases radiocarbon dates and other activity such as pits and structural remains were found. In only four cases Neolithic chambered tombs were noted. These were at Glecknabae on Bute in Scotland, Ballydorn in Northern Ireland and St Mary's Port/Alfred Pier and Dalby Mountain on the Isle of Man. Additionally at the site of Mayburgh Henge Mesolithic blades were found on the western edge of the henge.

These statistics and the exceptional nature of midden sites in the area indicate that the notion of Mesolithic middens representing precursors to megaliths, and the view of middens as the "go to" sites to understand the transition is problematic. Instead, to come to a clearer understanding of the transition what is needed is a return to the Mesolithic record, which is comprised predominantly of lithic scatter sites, and to consider these in a more nuanced, and socially situated way. By doing this and by coming to a greater understanding of some of the social dimensions of everyday life in this area this will then enable archaeologists to develop an account of the transition that addresses the issues that I have raised here – an account which challenges the socially impoverished narratives that exist and that enables us to move beyond the few exceptional sites to think about the nature of the transition in every day life.

Challenging the problems

To do this I suggest that one of the most helpful approaches to re-examine the Mesolithic in a more socially situated manner is to take a phenomenological approach to materials. In particular I think there are two ideas that are valuable. I find Ingold's (2007; 2011) views on material culture particularly useful. Here Ingold argues that the material word is always a world-in-formation, a messy tangle of material things continually engaged in the current of the lifeworld. Thus,

> "...the properties of materials ... cannot be identified as fixed, essential attributes of things, but are rather processual and relational. They are neither objectively determined nor subjectively imagined but practically experienced. In that sense, every property is a condensed story. To describe the properties of materials is to tell the stories of what happens to them as they flow, mix and mutate." (Ingold 2007, 14)

This view of how people, things and places emerge and continually emerge and re-emerge simply from the nature of their very being in the world is compelling. But to be able to say more about the past, I suggest that a second perspective that incorporates these ideas but that forms a more practical heuristic framework for thinking about the relationships between people, things and places in the past is Heideggers' work on background, disclosure and the equipmental totality.

In brief, in Being and Time, Heidegger introduces the notion of the background which, to all intents and purposes, is what we might now call the habitus. But where Bourdieu identified habitus as the tacit and habituated social rules that structure our being (Bourdieu 1977), for Heidegger the notion of background is more than this because it also incorporates the totality of material involvements of being in the world. In this way Heidegger (1962) sees everything connected in an equipmental totality. Much like Ingold's work, this is the notion that, in very simplistic terms, nothing exists in isolation – all material things exist in a referential totality. And it is through this referential totality that the world is disclosed. Moreover everything in the world shows up to us and we can make sense of it because it is revealed against a background. That background may be tacit and implicit and habituated but ultimately it enables us to understand what things, and people, and animals are, and what they are related and connected to, or not related and connected to, because they are never singular – they are always already disclosed as part of the equipmental totality. So we know a microlith is a microlith because it isn't an arrowhead, it isn't a hand axe, it isn't a blade of grass, it isn't a plate of jelly – critically we know it is none of these things, but we can only know this because of this totality of things that it exists in and the background against which it shows up. Alternatively Heidegger expresses this notion in more positive terms:

"... there is no such thing as an equipment. To the Being of any equipment there always belongs a totality of equipment, in which it can be this equipment that it is. Equipment is essentially 'something in-order-to' ... Equipment – in accordance with its equipmentality - always is in terms of it belonging to other equipment: ink-stand, pen, paper, blotting pad, table, lamp, furniture, windows, doors, room." (Heidegger 1962, 97)

Critically, things can never be understood as singular, and standing alone with an essential nature (Dreyfus 1992, 179). Instead things can only be disclosed to us through their relationships with other equipment (Thomas 1996), with other people and with the world (Cobb 2008).

I find this a useful way to begin to think about the material record from prehistoric hunter gatherer societies because the implication of this notion is that whether we are talking about a tiny stone tool, an occupation site or a whole landscape, it is through these things, which are all things that we can access as archaeologists, that we can examine how the world was disclosed, revealed and understood to prehistoric hunter gatherers. Of course I am not suggesting that we simply need to pick up a microlith, for example, and immediately we can understand it in its equipmental totality, as it was understood in the Mesolithic - far from it. Our understandings of things, places and people, whether we are talking about now or in the Mesolithic, are contextual and political. In fact Heidegger addresses this by demonstrating that whilst equipment is understood in relation to an equipmental totality or background, such a background only makes sense, and things are consequently only able to be disclosed to us as intelligible against this precisely because of the temporality, historicity and relationality of both background and disclosure (Heidegger 1962, 343-344; Thomas 2004b, 217-219; Wrathmall 2000). So what this means, then, is we have to look to the wider contexts of the material record to examine how materials might have acted to disclose and reveal the world in the Mesolithic period.

Thinking through things in the Mesolithic

The kind of approach outlined here enables us to think through things to come to a more socially situated and nuanced interpretation of the Mesolithic sites in the northern Irish Sea Basin through which people lived their daily lives. We can, for example, consider a single site such as the extensive spread of Mesolithic material around Gallow Hill (Donnelly and Macgregor 2005) and the nearby Littlehill Bridge (Macgregor and Donnelly 2001) on the Ayrshire coast.

Here, on the edge of a hill, on the raised beach, to the north of the once lagoonal and estuarine area at the mouth of the Water of Girvan (Donnelly and Macgregor 2005), both fieldwalking and more targeted excavation work have revealed extensive surface scatters of Mesolithic material, a series of mixed, unstratified Mesolithic deposits, in situ scattered lithic material and open site activity including pits,

hearths, areas of burning, stake holes and, several sub-oval, shallow sided scoops (MacGregor and Donnelly 2001, 5). Radiocarbon dates and the accumulations of material in this area, extending over approximately a square half kilometre, suggest that it was potentially revisited over a period of at least 1500 years in the late Mesolithic and the excavators have pointed to a focus on specialised blade and microlith production (Donnelley and MacGregor 2005, 58), and the repair of microlithic tools at the site (Donnelley and MacGregor 2005, 56).

With this in mind then, we can consider how the use of the site in daily life, and the activities of tool making and mending, and the repeated use of the place in general and the repeated re-use of specific areas for different types of occupation all worked to disclose understandings of the world. Here I argue that landscape setting may have played an important role in such disclosures. For example a wide range of raw materials from a wide range of sources, such as Pitchstone from the Isle of Arran, were being brought to and worked at the site (Donnelley and MacGregor 2005, 50). These materials in themselves would have acted as visceral reminders of journeys, disclosing an understanding of places and people across, entwined with and connected by the sea, or through the valleys through which they had been brought. In addition the location of site and the visual connections it affords would have worked to disclose the activities of the site against a background of further connections. For instance there are superb views from the site both inland over the Midland Valley, and out to the Firth of Clyde, the northern Irish Sea, the islands of Arran and Ailsa Craig (which can clearly be seen from the Antrim Coast) and much of Argyll and Bute, as well as the edges of the Southern Uplands. We could regard this site as a hub then where a series of material and visual connections across the land and the sea and across time too, were disclosed against the rhythms of daily life, whilst at once being entwined into the making, doing and being through which people's identities and understandings of the world were performed and negotiated.

Connections can also be drawn between landscape, materials and places as they all worked to disclose life in the Mesolithic on a broader scale too. For instance, of the sites visited in the northern Irish Sea basin as part of my doctoral research, there appeared to be a greater trend towards the restriction of views at later Mesolithic sites. This higher amount of visual restriction in the later Mesolithic corresponds with an increase in types of sites which emphasised a "hidden-ness" in their location, such as rockshelters (Figure 2), and thus would have worked to substantially restrict and direct the view from sites. If we consider how visual connections acted to disclose certain connections to places and therefore certain understandings of the world, then conversely we can suggest that by choosing locales where the view was restricted or focussed, later Mesolithic people were able to make a very specific statements about what sites revealed and the way in which certain relational connections could be made and disclosed.

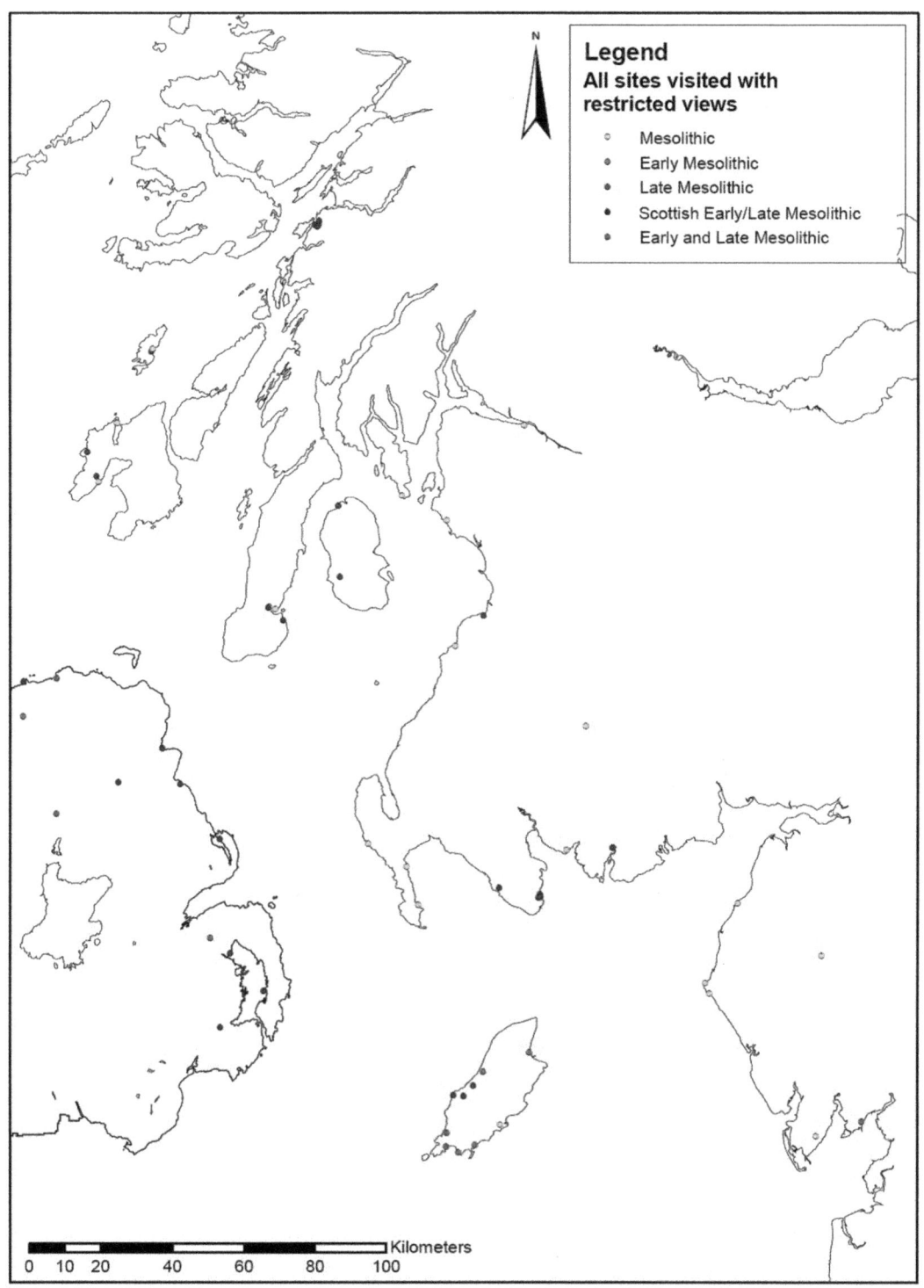

Figure 2: Sites visited in the study area with restricted views (Source: Cobb 2008)

Moreover if we consider not just landscape setting but also the tools and materials at these sites, there was a clear trend for these sites to have quartz present, and very few had microliths present. If we examine these materials and their symbolic values this helps to build a clearer picture of what may have been disclosed or revealed. Microliths, for instance, are often discussed in terms of their ability to entwine a wide range of people and things because of their composite nature (e.g. Finlay 2003; 2006). Quartz is also often discussed as perhaps having qualities that reflect important bodily substances (e.g. Fowler and Cummings 2003; Fowler 2004). Thus the absence of a tool type that combined multiple parts of the world and engendered the multiple authorship of identity, and instead the presence of a material that may have had a symbolic value in the mediation of other types of identity may be highly significant in trying to understand how late Mesolithic ontologies were disclosed at certain sites.

If we combine these factors then, and consider both landscape setting and the materials themselves such patterns may indicate that in the later Mesolithic these places worked in a very specific way, but in a way that contrasts quite markedly to open and unfettered sites such as the Oronsay middens, to root people to and transform them against, wider relational understandings of personhood. This ultimately lay in the very specific disclosure of the world that such places enabled through the combination of their visual affordances, and the related sets of material objects and practices that took place at them.

Conclusions

I have drawn upon two different but brief examples here, at different analytical scales, from a specific site to wider patterns, and I have demonstrated that there is a wide scope to consider the relationship between Mesolithic material culture and it's landscape setting in a way that can take us beyond the problematic limitations of current studies that I discussed in the first part of this paper. Drawing upon this idea of disclosure we can explore the very specific ways that people and things were disclosed against the context of the place in which they were used. And conversely we can examine how places and whole landscapes emerged and were themselves disclosed and revealed through the complicated interplay between people, things and places.

Of course this does not rule out using this approach to examine the exceptional, well preserved shell middens of western Scotland. But critically this is a heuristic framework which we can apply to all sites, the majority of which are lithic scatter and find spot sites (Cobb 2008). As a result we can begin to develop a much more representative, but none the less nuanced, and socially situated account of the Mesolithic. In turn, by building such narratives of the late Mesolithic, we can then build a more balanced view of the transition in the area. Rather than seeing a distinct change from socially impoverished, insular Mesolithic lifeways to a rich, social and ritual Neolithic world, instead a different view opens up. Acknowledging and examining how Mesolithic people were empowered and able to make active decisions in the way that they used materials, places and the landscape to disclose specific worldviews is a starting point. From this we can then begin to envisage a transition in which indigenous Mesolithic communities made choices about how to change, and about what elements of the Neolithic repertoire to adopt (Thomas 2003). With exciting new research demonstrating Neolithic activity in Britain as early as 4100 BC (Whittle et al 2011), we may therefore envision the transition not as a problematic disjunction in lifeways, but one in which pasts, future, new materials and old were worked into the disclosure of identities at the scale of daily life.

References Cited

Bevan, L., and Moore, J., (eds). 2003. *Peopling the Mesolithic in a Northern Environment*. BAR International Series 1157. Oxford, BAR Publishing.

Bourdieu, P. 1977. *Outline of a theory of practice*. Cambridge, Cambridge University Press.

Bradley, R. 1984. *The Social Foundations of Prehistoric Britain*. Harlow, Longman Group Ltd.

Cobb, H., L. 2007. Media for Movement and Making the World: Exploring Materials and Identity in the Mesolithic of the Northern Irish Sea Basin. *Internet Archaeology* 22 (Mesolithic Archaeology theme) http://intarch.ac.uk/journal/issue22/cobb_toc.html

Cobb, H., L. 2008. *Media for movement and making the world: An examination of the Mesolithic experience of the world and the Mesolithic to Neolithic transition in the Northern Irish Sea Basin*. Unpublished PhD Thesis, Manchester, School of Arts, Histories and Cultures, University of Manchester.

Cobb, H., L. 2009a. Being-in-the-(Mesolithic) world: Place, substance and person in the Mesolithic of Western Scotland. In S. McCartan, R. Schulting, G. Warren and P. Woodman (eds). *Mesolithic Horizons: Papers Presented at the 7th International Conference in the Mesolithic in Europe (Belfast 2005)*, 368-372. Oxford, Oxbow Books.

Cobb, H., L. 2009b. Tasks, Transformations, and Transitions: The transition from hunting and gathering to farming in the northern Irish Sea basin. In H. Glorstad, and C. Prescott (eds), Neolithisation as if history mattered, pp. 65-84. Lindome: Bricoleur Press.

Cobb, H., L., Coward, F., Grimshaw, L., and Price, S. (eds) 2005. *Investigating Prehistoric Hunter-Gatherer Identities in Palaeolithic and Mesolithic Europe*. British Archaeological Reports (International Series 1411). Oxford, BAR Publishing.

Conneller, C. (ed) 2000. New Approaches to the Palaeolithic and Mesolithic. *Archaeological Review from Cambridge* 17(1) 1239-150.

Conneller, C. 2004. Becoming deer. Corporeal transformations at Star Carr. *Archaeological Dialogues* 11(1), 37-56.

Conneller, C. 2011. *An Archaeology of Materials: Substantial Transformations in Early Prehistoric Europe*. London, Routledge.

Conneller, C., and Warren, G., (eds.) 2006. *Mesolithic Britain and Ireland: New Approaches*. Stroud, Tempus.

Cummings, V. 2009. *A view from the west: the Neolithic of the Irish Sea zone*. Oxford, Oxbow.

Cummings, V., and Fowler, C., (eds.) 2004. *The Neolithic of the Irish Sea : materiality and traditions of practice*. Oxford, Oxbow.

Donnelly, M., and MacGregor, G. 2005. The excavation of Mesolithic activity, Neolithic and Bronze Age Burnt Mounds and Roman-British Ring Groove Houses at Gallow Hill, Girvan. *Scottish Archaeological Journal* 27(1), 31-69.

Dreyfus, H. 1992. Heidegger's History of the Being of Equipment. In H. Dreyfus and H. Hall (eds). *Heidegger: A critical reader,* 173-185. Oxford, Blackwell Publishers.

Finlay, N. 2003. Microliths and Multiple Authorship. Mesolithic on the Move. In L. Larsson, H. Kindgren, K. Knutsson, D. Loeffler and A. Akerlund (eds.). *Papers Presented at the 6th International Conference in the Mesolithic in Europe, Stockholm, 2000,* 169-176. Oxford, Oxbow.

Finlay, N. 2006. Gender and Personhood. In C. Conneller, and G. Warren, (eds). *Mesolithic Britain and Ireland: New Approaches,* 35-60. Stroud, Tempus.

Heidegger, M. (1962 (1927)). *Being and Time.* Oxford, Blackwell.

Fowler, C. 2004. In touch with the past? Monuments, bodies and the sacred in the Manx Neolithic and beyond. In V. Cummings and C. Fowler (eds) *The Neolithic of the Irish Sea: Materiality and traditions of practice,* 91-102. Oxford, Oxbow.

Fowler, C. and Cummings, V. 2003. Places of transformation: Building Monuments from Water and Stone in the Neolithic of the Irish Sea. *Journal of the Royal Anthropological Institute (New Series)* 9, 1-20.

Hodder, I. 1982. *Symbols in action : ethnoarchaeological studies of material culture.* Cambridge, Cambridge University Press.

Ingold, T. 2007. "Materials against Materiality." *Archaeological Dialogues* 14, 1-16.

Ingold, T. 2011. *Being Alive: Essays on Movement, Knowledge and Description.* London, Routledge.

MacGregor, G., and Donnelly, M. 2001. A Mesolithic scatter from Littlehill Bridge, Girvan, Ayrshire. *Scottish Archaeological Journal* 23(1), 1-14.

Rowley-Conwy, P. 2004. How the West was lost: A reconsideration of agricultural origins in Britain, Ireland, and Southern Scandinavia. *Current Anthropology* 45, S83-S113.

Schulting, R., J. 2004. An Irish Sea Change: Some implications for the Mesolithic-Neolithic transition. In V. Cummings and C. Fowler (eds) *The Neolithic of the Irish Sea: materiality and traditions of practice,* 22-28. Oxford, Oxbow.

Sheridan, A. 2003. French connections I: spreading the marmites thinly. In I. Armit, E. Murphy, E. Nelis and P. Simpson (eds), *Neolithic settlement in Ireland and Western Britain,* 3-17. Oxford, Oxbow.

Sheridan, A. 2004. Neolithic connections along and across the Irish Sea. In V. Cummings and C. Fowler (eds) *The Neolithic of the Irish Sea: materiality and traditions of practice,* 9-21. Oxford, Oxbow.

Sheridan, A. 2007. From Pickardie to Pickering and Pencraig Hill? New information on the 'Carinated bowl Neolithic' in northern Britain. In A. Whittle and V. Cummings (eds), *Going Over: The Mesolithic-Neolithic Transition in North-West Europe,* 441-492. Oxford, Oxford University Press.

Sheridan, A. 2010. The Neolithisation of Britain and Ireland: the 'big picture'. In B. Finlayson and G. M. Warren (eds), *Landscapes in transition,* 89-105. Oxford and London, Oxbow Books and Council for British Research in the Levant (Levant Supplementary Series 8).

Thomas, J. 1996. *Time, Culture and Identity.* London, Routledge.

Thomas, J. 2003. Thoughts on the "Repacked" Neolithic Revolution. *Antiquity* 77(295), 67-74.

Thomas, J. 2004a. Current Debates on the Mesolithic-Neolithic Transition in Britain and Ireland. *Documenta Praehistorica* XXXI, 113-130.

Thomas, J. 2004b. *Archaeology and Modernity.* London, Routledge.

Thomas, J. 2007. Mesolithic-Neolithic transitions in Britain: from essence to inhabitation. In A. Whittle and V. Cummings (eds). *Going Over: The Mesolithic-Neolithic Transition in North-West Europe,* 423-439. Oxford, Oxford University Press.

Thomas, J. 2008. The Mesolithic-Neolithic Transition in Britain. In J. Pollard (ed), *Prehistoric Britain,* 58-89. Oxford, Blackwell Publishing.

Whittle, A., Bayliss, A. and Healy, F. 2011. *Gathering Time. Dating the Early Neolithic Enclosures of Southern Britain and Ireland.* Oxford, Oxbow Books.

Woodman, P., C. 2004. Some Problems and Perspectives: Reviewing Aspects of the Mesolithic Period in Ireland. In A. Saville. (ed), *Mesolithic Scotland and its neighbours: The Early Holocene prehistory of Scotland, its British and Irish context, and some Northern European Perspectives,* 285-297. Edinburgh, Society of Antiquaries of Scotland.

Wrathmall, M. 2000. Background practices, capacities, and Heideggerian disclosure. In M. Wrathmal, and J. Malpas (eds), *Heidegger, coping and cognitive science,* 93-114. Cambridge, Mass., MIT Press.

Young, R., (ed.) 2000. *Mesolithic Lifeways: Current Research from Britain and Ireland.* Leicester Archaeology Monographs 7. Leicester, University of Leicester.

Chapter 9
The Life of a Zebra Crossing
Biographical Approaches to Place

Ludvig Papmehl-Dufay

post-doc research fellow,
Linnaeus University,
Kalmar, Sweden

Abstract: *This paper discusses biographical approaches to place. The point of departure is the zebra crossing, a highly ordinary and anonymous kind of place taking part in the daily lives of thousands of people but generally not evoking special feelings or memories. However, single events can have a profound influence on the continued biography of the crossing, and it is argued that the same principle goes for archaeological places and monuments, too. Thus, when studying the biographical history of e.g. a passage grave, we should recognise and appreciate single events and their potential importance for the following development of the site's biography.*

Keywords: *creation of place, events, biographic history, memories*

Introduction

This paper explores the concept of place biography in archaeology. Place is obviously a concept at the core of the discipline, since all narratives and memories of the past have spatial references (Rubertone 2008, 16). Space and time are interdependent, and represent the two major limiting factors on people's movement and activities (Hägerstrand 1970). With his famous time geography model, Swedish geographer Torsten Hägerstrand illustrated this in a three-dimensional coordinate system where time constitutes the third (vertical) dimension added to a two-dimensional spatial plane (fig. 1). Individuals are represented in the model by points in the two-dimensional plane, and their movement through space and time are illustrated as lines straight, sloping or turning. A place could be illustrated in such a model as a straight vertical tube, through which individual actors pass by or spend some time, and perhaps carry out activities of various kinds. In this way, it is possible to graphically illustrate the long-term history of a place in terms of who visited it, when and for how long etc. In the context of this paper, the most interesting issue is: How is a place affected in this process?

Material biographies

The concept of biography in archaeology has most often been used in connection to portable objects (e.g. Appadurai 1986; Kopytoff 1986; Holtorf 2002). Focusing on the social lives of material things, the concept of artefact biographies aims at detecting and understanding changes in meaning that emerges from social action around and with the object, be it in the past, present, or both (Gosden and Marshall 1999,169). A central argument is that the life history of an object is a key to its social function. This function may vary

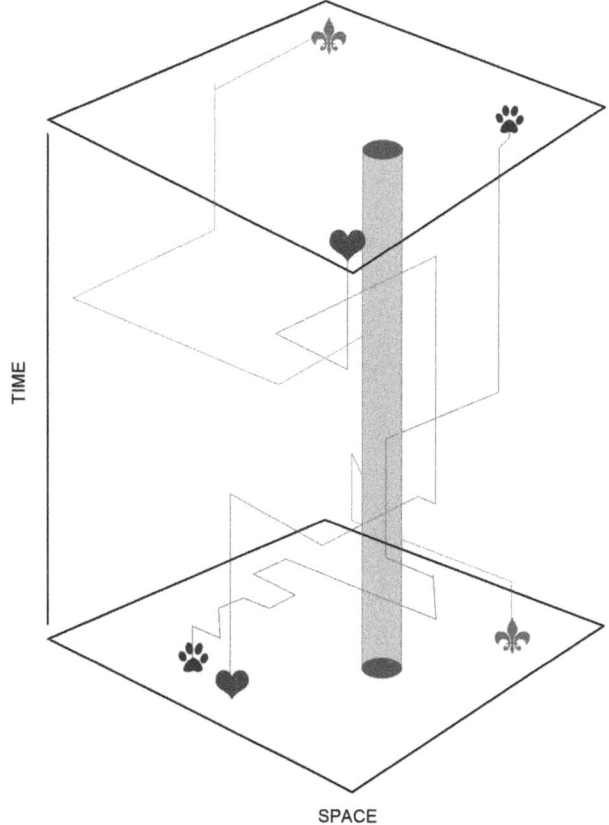

Figure 1. Simplified sketch of Hägerstrand's time-geography model, with an addition of place illustrated as a straight vertical tube. Drawing by the author.

considerably over time, and that this in turn is vital for an understanding of its practical dimensions as well. This is relevant to archaeology: artefacts that we find can only be understood through an appreciation of their social context, where communication and negotiation of meaning are central. Thus, to be able to understand anything about the social role of things, we must study their history and their changing roles in various social contexts in which they have participated. Material objects in this perspective should be regarded as actors, capable of taking part in communication with other material and human actors (LaMotta and Schiffer 2001).

The participation of material objects in social interaction and communication is still a highly active process in western society today. One need not turn to the Kula exchange system in the Trobrianders of Papua New Guinea to find examples of this; countless cases can be drawn upon from modern western society where objects are valued and treasured based on their specific biographical histories. Key concepts here are memory and remembrance; the life history of an object may involve specific events and paths that render the object a certain status, but it is only as long as these events are remembered and acknowledged that the status is retained. A black plastic pen used by Iron Maiden drummer Nicko McBrain to sign a few autographs at a rock concert in 1991 may still be treasured today, based solely on the (remembered or just claimed) fact that the famous drummer held it in his hand for 30 seconds that day some 20 years ago. This event is not in any clear way detectable today; the pen itself does not show any physical traces of it. It is solely through the remembrance of this particular event within a specific social context that the plastic pen becomes a valuable, and it is only through keeping this memory alive that the artefact keeps its very specific life history. Antique experts call this the "provenance" of the item, its documented and relevant history. The actual accuracy of this claimed history might be of less importance, as long as the social actors involved hold it as true and relevant.

Of course, this process is not in any way restricted to portable or personal objects. Larger objects, such as the individual sarsens in the enormous henge monument at Avebury, will have a life history of their own and may be approached in a biographical manner as well (Gillings and Pollard 1999). Megalithic monuments may seem durable and static, but their meaning is by no means carved in stone (Gillings and Pollard 1999:185). In the same way that the history of an object is read in its wear (Jones 2007, 21), the history of a place is recorded through traces of events and activities in connection to the site. In the case of Avebury, the monument and all of its parts have an intriguing history that, at least in some cases, can be archaeologically detected and followed right from the late Neolithic through prehistoric and medieval and up to present times (Gillings and Pollard 1999). In this case, the life history is "remembered" (or rather, re-remembered) through archaeology, i.e. events long forgotten are brought to light once more by means of archaeological investigation

and reasoning. Thus, rather than simply representing dead fossil traces of a distant past, prehistoric monuments can be seen as living and experiencing actors with a rich and varied past, an active present and an unforeseen future (Holtorf 1998). An excavation of a passage grave in this sense is just another (not necessarily the last) event in the life history of that particular monument.

The life of a Zebra Crossing

From this follows that the biography of a place is continuously being determined by itself and by the actors involved. Acts and events performed in connection to the place will have some form of impact on the course of following events, and so it continues. Some events may not have such a great impact on the history of a place in the long run, while other major or seemingly simple events can change it completely. Typically, acts and performances following the expected and previously dominating pattern at a site will serve to maintain and gradually develop the history of the place, while events deviating from the expected pattern can result in drastic changes in the life history of the place. An illustration of this is provided by a certain zebra crossing at Abbey Road in London (Kruse 2005:94ff): this is in essence a highly ordinary and anonymous place in modern terms, taking part in peoples' life on a daily basis but generally not the kind of place that evokes special feelings or deserves particular remembrance. The act of people walking such a crossing should not generally be expected to profoundly change the life history of the place. However, today this crossing is a major tourist attraction, and thousands of people come here every year just to see and experience the place, to cross the road at the crossing and take a picture while doing so (fig. 2). Actually, there is even a live web-cam of the place, recording activities 24 hours of the day (http://www.abbeyroad.com/visit/). The life history of the zebra crossing at Abbey Road was completely revolutionized on August 8th 1969, when a certain rock band crossed the road here and Scottish photographer Iain Macmillan took a few pictures of the event. What happened? The place as we see it today existed several years prior to 1969, and apparently it has not undergone any major physical alterations since. If it were not for the photographs, and the fact that one of them was used as a front cover for the last album to be recorded by The Beatles, the event of the band walking the crossing would probably not have altered the biography of the place in the long term and it would certainly not have been widely remembered and celebrated today. The event as such, the band walking the crossing, was not deviating from the "normal" activities at the place, but the way it was communicated sure was. Apart from having been immortalised as a piece of rock n' roll history, the event has triggered an enormous amount of quite extraordinary scenes at the site, such as the almost naked members of the rock band Red Hot Chili Peppers re-enacting and parodying the famous album cover ("The Abbey Road E.P."). The site is still an ordinary urban place, the intersection between Abbey Road and Grove End Road, and it takes part in

Figure 2. The zebra crossing at Abbey Road. Photo: Private.

vernacular daily life continuously, while at the same time it has the role of a sacred place of pilgrimage for fans of The Beatles around the world.

Thus, history and memory are vital to the development of the biography of a place. Single events may change dramatically the course of following events, and the meaning of the place may become fundamentally altered along the way (see Gosden & Marshall 1999: 169; Gillings & Pollard 1999). Countless other examples of this same phenomenon can be drawn from various religious places, where the claimed memory of divine appearances sometimes results in the most fascinating developments of place biographies. At Fatima, Portugal, the claimed apparition of the Virgin Mary to three children in 1917 resulted in a major religions centre still in use today. In the decade following the original apparition, an estimated 2 million pilgrims visited the site, and today millions of people still visit every year. In this case, the triggering event must be regarded as highly unusual and unexpected. The resulting change of the place was so radical (from an olive tree in the middle of nowhere to an urban religions centre) that one should perhaps regard it as an entirely new place not existing prior to May 1917.

The meaning of places

There are, in my view, no reasons to believe that the process outlined above should not be applicable to the biographies of ancient sites and monuments as well. Even though not all strange ancient places should be interpreted as holy sites of pilgrimage, previous events at a site and their remembrance should not be neglected. Places of death and mourning are especially well suited to illustrate the point. Today the meaning of an Iron Age graveyard lies in our knowledge of the history of the site, drawn from archaeology, local tradition and/or mythology. In this way many graveyards are held by local enthusiasts to be immensely important locations, worthy of the deepest respect and awe by those visiting such a place today. It provides a link to the past and a key to people's distant ancestry, and thus it is important in the process of creating and maintaining a local identity. The locus of the graveyard, the site itself, serves as a metaphor for ancient roots and belonging (Eriksen 1996). Any alteration of the physical appearance of the site (be it clearance of vegetation or removal of stones) has the potential of altering the meaning of the place, as it is perceived today. However, even though we see the meaning of the place in this case as deeply historically embedded, and even though we think we know roughly what this place meant to people in the past, our perception of it most likely differs completely from how it was perceived during the Iron Age. Memories and stories connected to the persons buried and the events that led to their death as well as mythical and religious beliefs connected to the various acts and practices performed at the grave yard, were all parts of the collective memory of such a place in prehistory. Thus, its meaning was completely defined by its biography.

Having established that acts and events performed at a site are crucial to its continued development as a place, how can this be detected archaeologically? How can the biography of a megalithic tomb be studied from the perspective outlined above? We can always speculate about the most amazing events taking place at a site in the past and thereby setting the trend for coming events, but in the end we are constrained to events that leave materially detectable traces. This leaves an enormous chunk of the place's history unreachable through archaeology: we can never establish what was said, sung or danced at a place at certain occasions, and we will never be able to fully reconstruct mythical stories of the divine aspects of various sites in the past. Still, the notion that the biography of place is constituted by earlier (real or claimed) events and their remembrance is of great importance to every archaeological study of long-term change or continuity of a site or a region. Also studies of a narrower time frame should benefit from a retrospective understanding of the place's history and social significance.

Understanding megalithic tombs in a long-term perspective

Megalithic tombs represent one of the earliest monumental grave structures to be built in Scandinavia, and the first real effort of using stone architecture to manifest something in the landscape. By their very nature, megalithic monuments use memory and references to the past to create a notion of infinity and value (Jones 2007, 21). Their statement is as much about the past as about the future (Holtorf 1998).

A massive body of research has dealt with various aspects of the megalithic tradition of the Scandinavian TRB culture, relating it to aspects such as religion, economy, demography, migration, etc. (e.g. Strömberg 1971; Lidén 1995; Persson and Sjögren 2001; Sjögren 2004; Ahlström 2009). The distribution of megalithic tombs is not identical to that of the TRB culture in the area, however, and the problem still remains to be solved as to why megalithic structures were extensively built in some areas and not in others (e.g. Hallgren 2008). The most obvious puzzle in this sense is perhaps the extremely dense concentration of passage graves in central Västergötland, Sweden, consisting of more than 250 tombs in the Falbygden region (Persson and Sjögren 2001; Sjögren 2004).

On the island of Öland, off the Swedish SE coast in the Baltic Sea, an isolated group of four megaliths is located

in the parish of Resmo on the southwest part of the island (fig. 3). The three passage graves are located within an area of only a couple of hundred meters, while the dolmen is situated c. 2.5 km to the north (Papmehl-Dufay 2006). Viewed from the total distribution of megaliths in Scandinavia, the Öland megaliths represent an eastern periphery far away from the nearest concentration of tombs. The area is rich in Neolithic remains however, and the tombs mirror a relatively strong presence of the TRB culture in this area. In recent years several TRB sites have been excavated on the island, the results indicating cultural links with (megalithic) southwest Scandinavia as well as (non-megalithic) east central Sweden (Papmehl-Dufay 2009; Alexandersson and Papmehl-Dufay 2010). Earlier research saw the Öland tombs as reflecting a migrating community from Scania or Denmark (i.e. southwest Scandinavia) who settled on Öland bringing their funeral traditions with them (Stenberger 1948). Recent research has been somewhat more ambivalent, not excluding migrations of people but also proclaiming the movement of objects and ideas (Papmehl-Dufay 2006; Linderholm 2008). Geological similarities between Öland and the Falbygden region in Västergötland have been discussed, and the insular landscape setting has also been suggested an important factor (Sjögren 2004; Papmehl-Dufay 2003).

How can we explain such an isolated concentration of megalithic tombs? So far only one of the Öland tombs has been excavated (Arne 1909), but most probably all four tombs were built within a relatively narrow timeframe in the late EN, around 3400 BC (Papmehl-Dufay forthcoming). This corresponds to the very initial phase of megalith building in south Scandinavia. Through its construction as a massive stone-built monument, the erection of a megalithic tomb in any location represents a major change of that place, physically and probably also symbolically. It is safe to assume that the biography of the place was more or less dramatically changed through the construction of a megalith. This is further emphasised by the fact that many tombs, among them the excavated Öland passage grave, were used as burial chambers periodically for several millennia after their initial construction (Eriksson *et al.* 2008; Papmehl-Dufay forthcoming). Still, in many cases the places of megalith construction probably already were loaded with memories and biographical history, and thus the erection of the tomb may not have been the original triggering event in changing the biography of that place. Is it possible that megalithic tombs in some cases were built as a result of some earlier extraordinary event at a site or in a region, or in response to some extraordinary place-bounded myths or memories? In some cases earlier monuments in the form of EN long barrows are present in the same areas as concentrations of megaliths, and a few examples even exist of megaliths having been built into earlier long mounds (e.g. Larsson 2002). In these cases, it seems that the rather specific direction of the place's biography towards special significance -connected to ancestors and memories - started before the megalithic tombs were built.

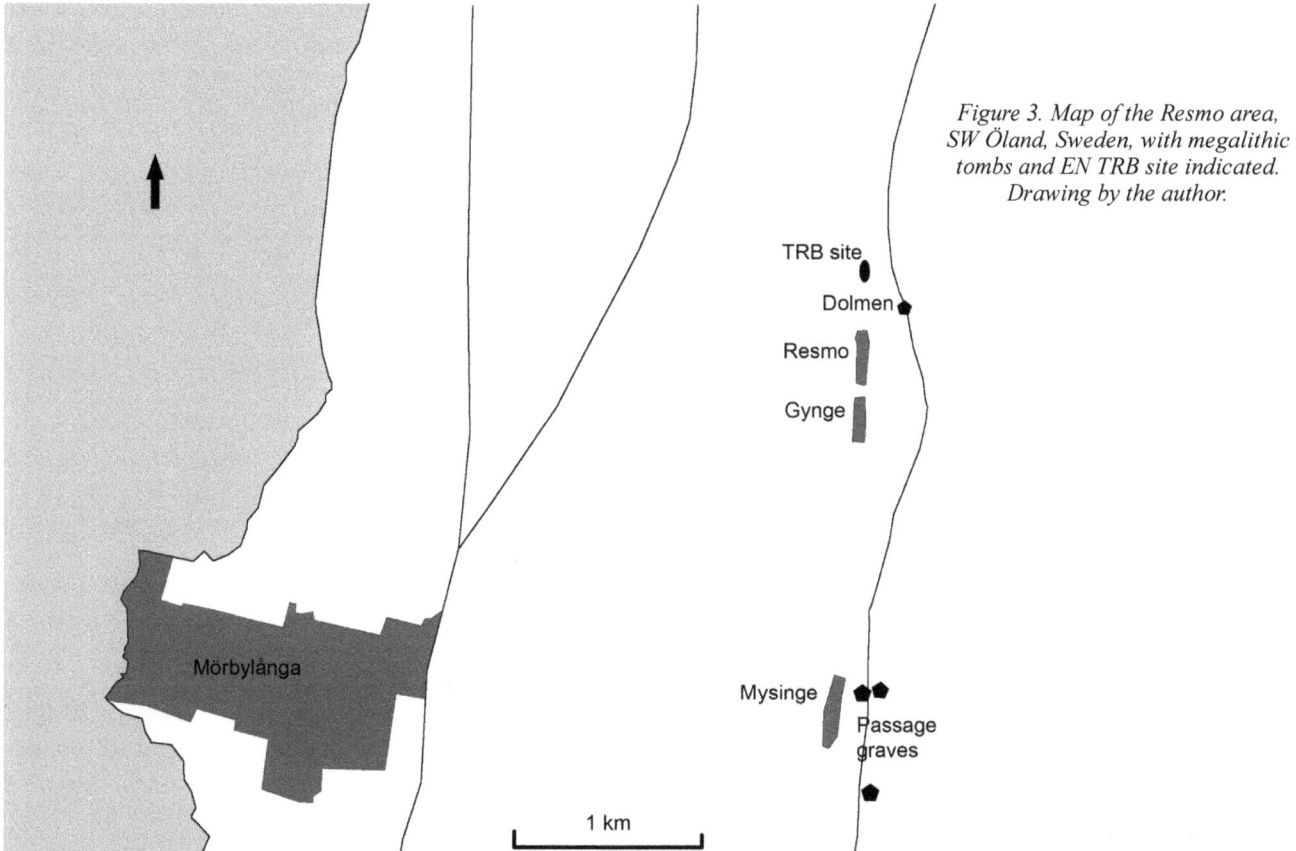

Figure 3. Map of the Resmo area, SW Öland, Sweden, with megalithic tombs and EN TRB site indicated. Drawing by the author.

Concerning the Öland tombs and their setting, we don't know too much about the period preceding their construction. Recently, however, a TRB site with large amounts of EN I pottery (c. 3900–3500 BC) was found and partly excavated only a few hundred meters from the northernmost of the tombs (Papmehl-Dufay 2009) (see fig. 3). A thick cultural layer rich in finds of pottery, flint and some burned bone was investigated, along with various sunken features such as pits and small postholes. No built structures could be identified, and it is still not clear what kind of activities the rich cultural layer represents.

The site is located at around 40–41 m above the present sea level, and occupies a monumental setting in the generally extremely flat Öland landscape. The majority of Neolithic settlement sites presently known on the island are located on the sandy beach ridges close to the Neolithic shoreline, around 10–15 m above the present sea level (Alexandersson and Papmehl-Dufay 2010), which makes the TRB site in Resmo stand out when concerning its landscape setting. The presence of the rich ENI cultural layer at Resmo, some 300 m from a location where a dolmen was built only a few generations later, clearly shows that this place was not neutral prior to megalith construction. The social context in which the pottery was used and discarded is still not clear, but from the sheer amounts and the great technological homogeneity (Papmehl-Dufay et al ms) it is suggested that the find represents some quite extraordinary kind of activities not connected to domestic dwelling in the first place.

Conclusion

My main argument in this paper is that the meaning of a place is completely made up of its history as remembered and perceived by the people involved, and that individual events may have a dramatic impact on the course of following events at the site. Precisely what these events are is not the main concern, but I think that an understanding of this phenomenon is crucial to the long-term studies of archaeological places. The zebra crossing at Abbey Road and the holy city of Fatima are but two examples of places with a quite extraordinary biography, where the triggering event would be impossible to trace archaeologically. In the case of Abbey Road, even the exceptional course of events following that photography session in 1969 and continuing up to this day would be difficult to detect and understand archaeologically. In Fatima we would see a city of great religions significance arising rather abruptly sometime in the early decades of the 20th century. Conceptually this case bears many similarities to prehistoric sites of religions importance, e.g. places of offering or cult.

The case of Öland briefly outlined above served to illustrate that the archaeological biography of a place is made up not only of what is or have been visible above ground. In the study of an area with megalithic tombs, it seems natural to regard the initial construction of monuments as a turning point in the place's biography. This is not necessarily so, however; we do not know what myths or memories where connected to the place before the monuments were constructed, and even if we find an earlier archaeological site representing activities connected to such meanings it is not easy to discern this archaeologically. The EN TRB site in Resmo clearly indicates that the place where the dolmen was erected a few hundred years later already had a history and was a part of people's memory.

References

Ahlström, T. 2009. *Underjordiska dödsriken. Humanosteologiska studier av neolitiska kollektivgravar.* Gothenburg, Coast to coast-book 18.

Alexandersson, K. and Papmehl-Dufay, L. 2010. *Två stenåldersboplatser i Runsbäck.* Unpublished excavation report, Kalmar läns museum arkeologisk rapport 2010:X.

Appadurai, A. (ed) 1986. *The social life of things. Commodities in cultural perspective.* Cambridge, Cambridge University Press.

Arne, T. J. 1909. Stenåldersundersökningar II. En Öländsk gånggrift. Fornvännen pp 86-95.

Eriksen, T. H. 1996. *Historia, myt och identitet.* Stockholm, Bonnier Alba.

Eriksson, G., Linderholm, A., Fornander, E., Kanstrup, M., Schoultz, P., Olofsson, H. and Lidén, K. 2008. Same island, different diet: Cultural evolution of food practice on Öland, Sweden, from the Mesolithic to the Roman period. *Journal of Anthropological Archaeology* 27, 520-543.

Gillings, M. and Pollard, J. 1999. Non-portable stone artefacts and contexts of meaning: the tale of Grey Wether (www.museums.ncl.ac.uk/Avebury/stone4.htm). *World Archaeology* 31(2), 179-193.

Gosden, C. and Marshall, Y. 1999. The cultural biography of objects. *World Archaeology* 31(2), 169-178.

Hägerstrand, T. 1970. What about people in regional science? *Papers of the Regional Science Association* 24, 1–12.

Hallgren, F. 2008. Identitet i praktik. Lokala, regional och överregionala sammanhang inom nordlig trattbägarkultur. Uppsala, Coast to coast-books 17.

Holtorf, C. 1998. The life-histories of megaliths in Mecklenburg-Vorpommern (Germany). *World Archaeology* 30(1), 23-38.

Holtorf, C. 2002. Notes on the life history of a pot sherd. *Journal of material culture* 7(1), 49-71.

Jones, A. 2007. *Memory and material culture: tracing the past in prehistoric Europe. Topics in contemporary archaeology.* Cambridge, Cambridge University Press.

Kopytoff, I. 1986. The cultural biography of things: commoditization as process. In A. Appadurai (ed), *The social life of things. Commodities in cultural perspective*, 64-91. Cambridge, Cambridge University Press.

Kruse, R. J. 2005. A cultural geography of The Beatles. Representing landscapes as musical texts (Strawberry Fields, Abbey Road and Penny Lane). Studies in popular culture 3. New York, the Edwin Mellen Press.

LaMotta, V. M. and Schiffer, M. B. 2001. Behavioural archaeology. In I. Hodder (ed) *Archaeological theory today*, 14–64. Cambridge, Cambridge University Press.

Larsson, L. 2002. Långhögar i ett samhällsperspektiv. In L. Larsson (ed) *Monumentala gravformer i det* äldsta *bondesamhället.* Report Series No 8.147–171. Lund, University of Lund, Department of Archaeology and Ancient History.

Lidén, K. 1995. Megaliths, agriculture, and social complexity. A diet study of two Swedish megalith populations. *Journal of Anthropological Archaeology* 14, 404–417.

Linderholm, A. 2008. Migration in prehistory. DNA and stable isotope analyses of Swedish skeletal material. *Theses and papers in scientific archaeology 10.* Stockholm.

Papmehl-Dufay, L. 2003. Stone Age island archaeology. In C. Samuelsson and N. Ytterberg (eds), Uniting Sea. Stone Age Societies in the Baltic Sea Region, OPIA 33, Uppsala: department of Archaeology and Ancient History, 180-203.

Papmehl-Dufay, L. 2006. Shaping an identity. Pitted Ware pottery and potters in southeast Sweden. Theses and papers in Scientific Archaeology 7. Stockholm, Stockhom University.

Papmehl-Dufay, L. 2009. En trattbägarlokal i Resmo. Arkeologisk förundersökning och särskild arkeologisk undersökning 2008, Resmo 1:13, 1:14, 1:15 och 1:16, Resmo socken, Mörbylånga kommun, Öland. Unpublished excavation report, Kalmar läns museum arkeologisk rapport 2009:29.

Papmehl-Dufay, L. In Press. The passage grave at Mysinge, Öland, SE Sweden in a long term perspective. To be published in *Archaeologica Lituana.*

Papmehl-Dufay, L., Stilborg, O., Lindahl, A. and Isaksson, S. *For everyday use and special occasions. A multi-analytical study of pottery from two early Neolithic TRB sites on the island of* Öland*, SE Sweden.*

Persson, P. & Sjögren, K.-G. 2001. *Falbygdens gånggrifter. Del 1. Undersökningar 1985-1998.* 34, Göteborg, GOTARC Series C.

Rubertone, P. E. 2008. Engaging monuments, memories, and archaeology. In P.E. Rubertone (ed), Archaeologies of place making. Monuments, memories, and engagement in Native North America, 13–33. Walnut Creek, Left Coast Press.

Sjögren, K.G. 2004. Megalithic tombs, ideology and society in Sweden. In H. Knutsson (ed) *Coast to coast – arrival. Results and reflections*, 157-182. Uppsala, Coast to coast-book 10.

Stenberger, M. 1948. Det forntida Öland. In B. Palm, L. Landin, and O. Nordman (eds), Öland. 1, 299-398. Lund, Lund University.

Strömberg, M. 1971. Die megalitgräber von Hagestad. Zur problematik von graubbauten und grabriten. *Acta Archaeologica Lundensia* Series in 8 (9). Lund, Lund University.

CHAPTER 10
FLINT AS A MEDIUM OF SOCIAL CHANGE

Kenneth Alexandersson

Kalmar County Museum

Abstract: *During late Mesolithic and early Neolithic the same major changes appears along the Kalmar coast as in South Scandinavia as a whole.*

During middle Mesolithic the settlement pattern In Kalmar sound area consists of small sites scattered in the landscape. A distinct change in the settlement pattern is seen during late Mesolithic, with a concentration of activities in to coastal areas near river estuaries and lagoons along the coast. These late Mesolithic sites play an important role in local networks and are often large with significant quantity of lithic material. The lithic materials show an inter-reaction between the local and the distant and consist of lithic raw materials from different geographical provenances.

These sites hold an important role in the social integration and spreading of new ideas and mentalities. Still lacks more detailed analysis of the lithic material which would improve our knowledge of local cultural developments.

Keywords: *South Scandinavia, Late Mesolithic, Early Neolithic, lithic material*

During the transition between late Mesolithic and early Neolithic, large scale social changes are visible in Southern Scandinavia; new nutritional resources, changing settlement patterns and a new set of material culture. The changes are not something that only occurs on a physical plane, but is equally a question about changes in people's mentality.

These changes occur concurrently over large areas. The introduction of the new material culture, as part of the neolithisation, is noticeable simultaneously in an area from Skåne in the south to the Mälardalen region in the north (Persson 1999). Obviously the changes did not occur uniform across this vast area. There are a variety of local forms and adaptations. In this article we shall see how some of these local manifestations appear in the Kalmar region in southeast Sweden during the late Mesolithic (fig 1).

Kalmarsund area

In the southern part of Kalmar County, on Öland and Möre (the area on the mainland around Kalmar), a large number of Stone Age settlement sites are known from an extensive chronological sequence, early Mesolithic to Late Neolithic. In this article, I will look more closely at the settlement pattern during the late Mesolithic and the transition to the early Neolithic. As a background for the settlement pattern, I will first give a short discussion of the geographical conditions in the area.

The bedrock along the coast of Möre and on Öland consists of sedimentary Ordovician rocks and differs from the rest of the region where igneous rock predominates. The bedrock in Möre are made up of Cambrian sandstones, which gives the area a flat coastal plain. The area is also characterized

by an almost complete absence of both larger and smaller lakes. The sediment in the area is dominated by, water eroded till and glaciofluvial sediments, as eskers run in a south-easterly direction through the landscape. During the Stone Age this has contributed to a relative large number of islands elongate the coast of the mainland. In Möre several small rivers flows across the landscape into Kalmarsund. On Öland the bedrock is made up of slate and limestone. The highest parts on Öland are located on the western edge of the island, and the bedrock is slightly tilted to the east. The western escarpment serves as an important watershed in the province; most waterways are running to the East Coast of the island. Along the coast of Öland there are only a scarce numbers of small islands.

Changing water levels

The shoreline displacement in south-eastern Sweden and in the Baltic Sea in general, shows a complex history. Knowledge about the different stages in the Ancylus and Littorina transgressions are essential within Stone Age research (Åkerlund 1996). The flat landscape of the region means that changes in water level has major implications for how the landscape has been shaped. The highest levels of these two transgressions differ between the southern and northern parts of the region. The Ancylus transgression reaches its highest level in the area around 8300 BC (fig 2). In the northernmost part of Möre the shoreline at that time is about 20 meters above the modern sea level, at the same time the sea level in the southern part of Möre is about 8 meters above the modern sea level. Consequently, today, contemporary coastal sites are situated at increasing levels to the north. The Littorina transgression approaches its highest level around 5400 BC and a more sustained

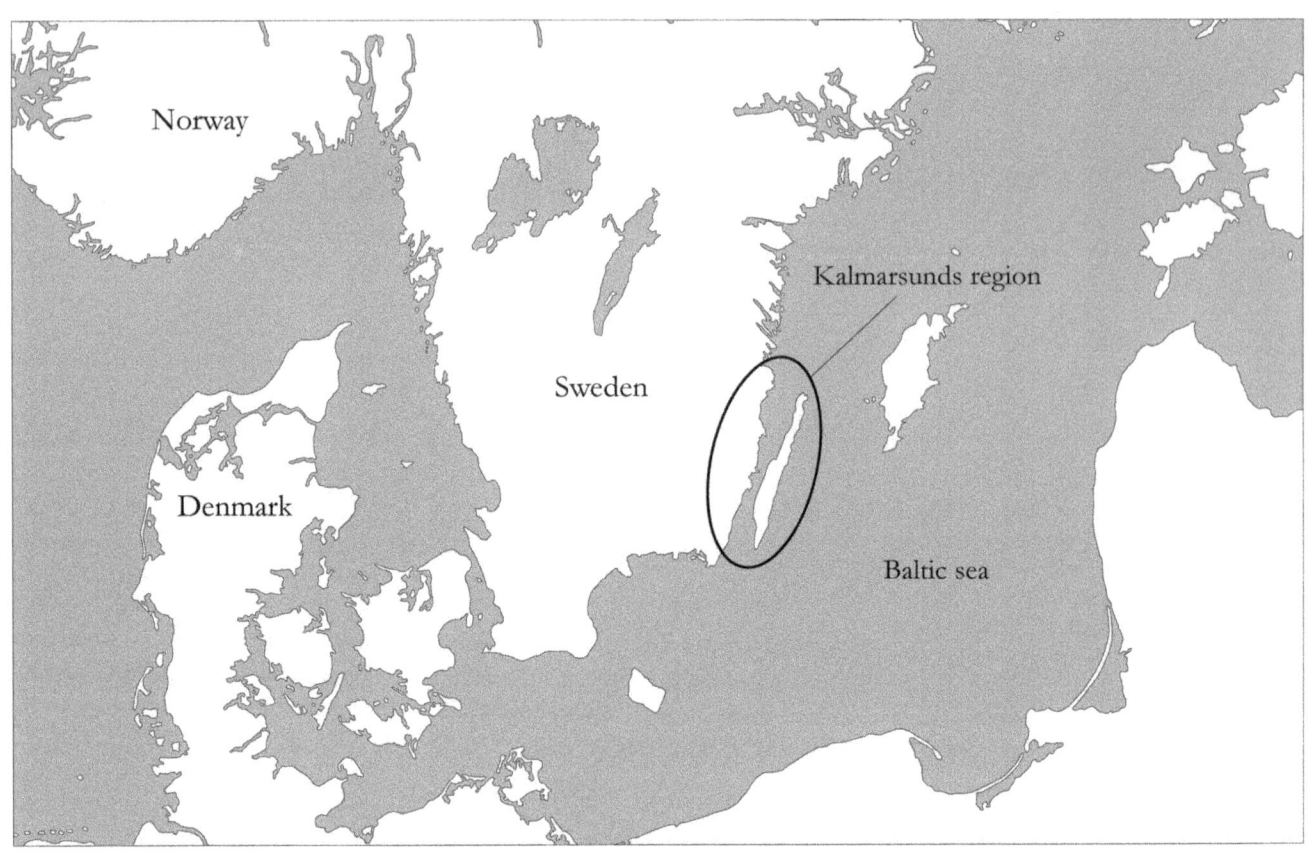

Fig 1 Map of Kalmarsund region

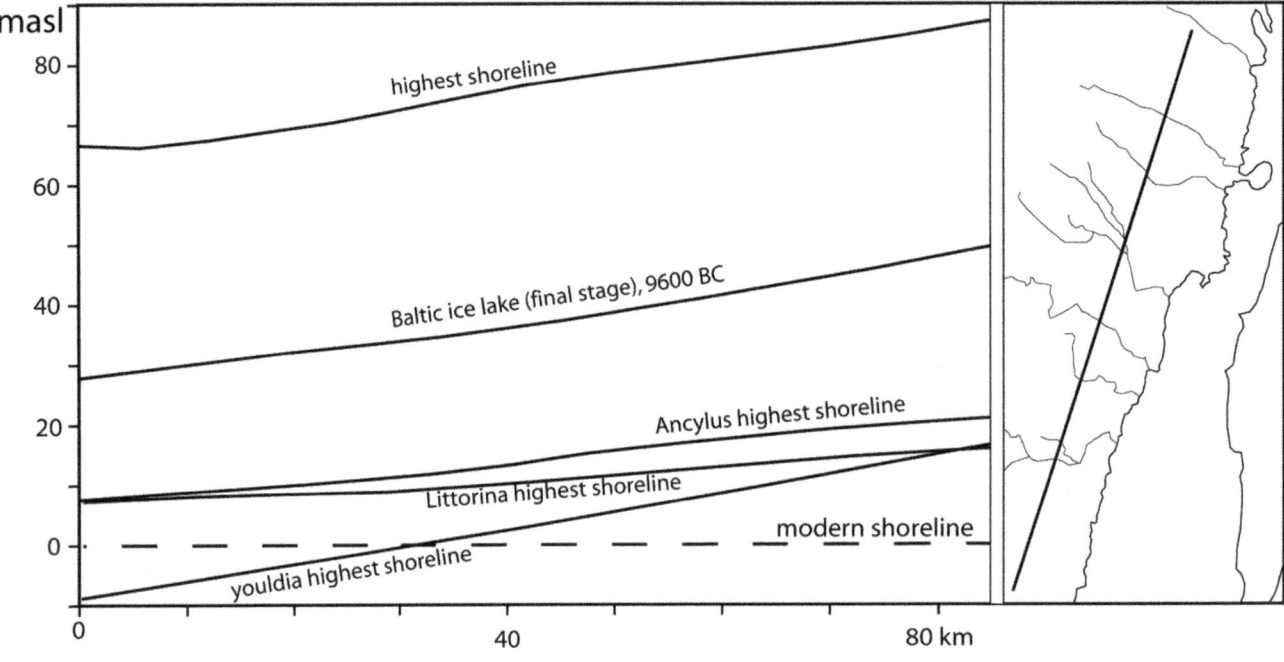

Fig 2 Diagram showing shoreline displacement in Möre (adapted after Svensson 2001).

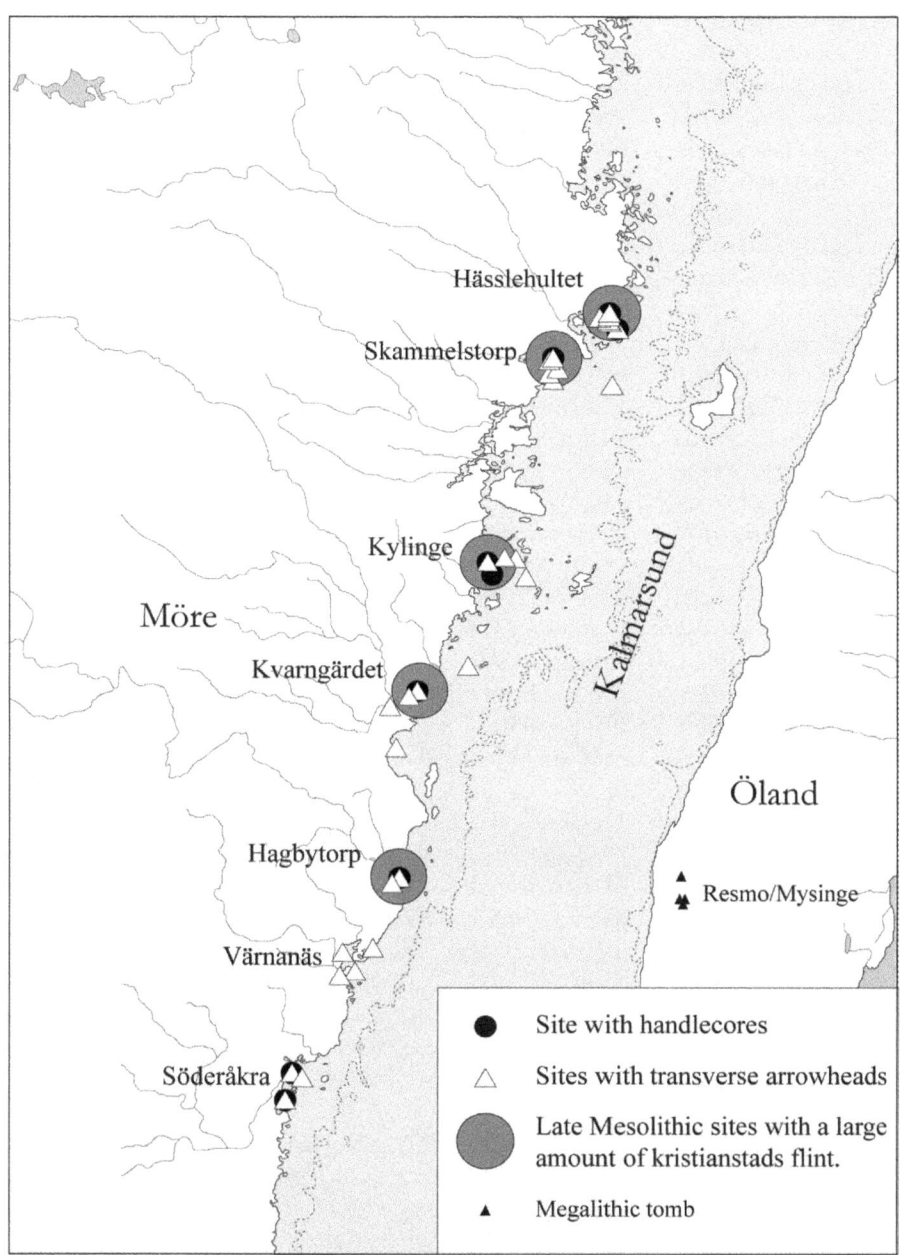

Fig 3 Late Mesolithic sites. The map shows the shoreline aprox 5000 BC. Dashed line shows the modern shoreline (adapted after Alexandersson 2001).

reduction in water level is visibly around 3900 BC. The Littorina transgressions highest level in the northern parts of Möre is about 16 meters above modern sea level. In the south of Möre it's around 8 meters above modern sea level (Svensson 2001). This means that the maximum level of Ancylus and Littorina transgression coincides in the southern part of Möre.

Different lithic raw materials

Another characteristic features for Stone Age societies in the region is the use of many different lithic raw materials for tool production (Alexandersson 2001). Kalmarsund area is located between quartz-dominated tool traditions in the north and flint-dominated in the south (Åkerlund 1996; Alexandersson 2007). Some of the lithic raw materials are of local origin, as quartz, porphyry and quartzite. A specialty for the region is the local Ordovician flint that occurs on the beaches of eastern Öland.

There is also a presence of lithic raw material from more distant sources. This reflect contacts with other regions, direct or indirect, far beyond the local areas, for example senonian, danian and kristianstad flint from southern Scandinavia.

A new settlement pattern

During the middle Mesolithic settlement patterns consist of relatively small sites, evenly dispersed in the landscape. It is difficult to single out sites that appear to dominate over the others. During the late Mesolithic, it is possible to see a distinct change in the distribution of settlement sites. It becomes clear that settlement is concentrated in specific areas of the landscape and activities increases near the mouths of rivers and lagoons along the coast (fig 3).

It is also possible to see a greater variation in the size between different sites within the settlement pattern. On the mainland, in Möre, there are at least five major sites spread along the coast, distinguished by a large amount of lithic waste; Hässlehultet, Skammelstorp, Kylingen, Kvarngärdet and Hagbytorp. These large sites are situated near the mouths of rivers or lagoons along the coast (Alexandersson 2001). There seems to be a clear link between river mouths and larger more frequently used sites. It's not possible to find the same distinctive settlement pattern or landscape use at the opposing west coast on Öland at the same time.

The lithic material on these sites is dominated by kristianstadsflinta and could be dated to a period stretching from late Mesolithic to early Neolithic. It's possible to see that activities on those sites have gone on for a long period of time. On the other hand some of the sites, for example Hässlehultet, are situated near a coastal lagoon. Lower water levels, as a consequence of the shoreline displacement has dried up the lagoon and resulted in people moving to new places. At the mouths of rivers, however, the basic elements remain. The river mouths importance for the settlements pattern could be seen during middle Neolithic. In some areas the settlements sites moved downstream, closer to the mouths of rivers; as an effect of the shoreline displacement these large sites near the mouths of rivers and lagoons seems to have a key role within the settlement pattern.

Lithic material

It's possible to interpret the large amount of lithic waste on some of the sites from different perspectives. A large amount of lithic waste could be an effect of a more stable settlement pattern. It could also reflect that certain activities are concentrated at the same place for long periods of time.

As mentioned earlier the Littorina transgression reached its highest peak in the area around 5200 BC. This means that the water level in Möre remained stable for a long time. During more than 1 000 years only minor fluctuations occurs in water level, around 1-2 meters (Svensson 2001). As a result locations can be assigned strong historical legitimacy.

The lithic stone materials on these large sites reflect production for everyday use, such as tool production. The lithic materials are dominated by non-local raw materials as different types of flint from the south. At the same time the material also shows features of ritualized activities, with burnt offerings and deposits. The phenomenon is familiar, being seen in other parts of southern Scandinavia and reflects the emergence of a more stable settlement pattern (Karsten 1994).

Hagbytorp

It's time to look closer at one of the larger late Mesolithic settlement sites, Hagbytorp. The site is located on a slight slope on the northern shore of the contemporary mouth of the Hagbyåns River. Today the area is used as arable land,

and the site was discovered in the 1960's when the land owner collected a large number of ground stone axes in connection with farming activities. Today there are a total of some 50 axes from the site. Morphologically the axes show large variation; ranging from late Mesolithic peaked ground stone axes to Neolithic ground stone axes (fig 4). Notably, polished flint axes are absent in the material.

In the 1980s, a private collector performed several surface collections at the site. This material has now been incorporated in the Kalmar County Museums collections and contain over 12 000 artifacts. In the surface collected material, there is a predominance of non-local raw materials as senonian-, danien- and kristianstads flint. The material shows intensive use and is spread over a large area. The typologically datable artifacts show that the site of Hagbytorp has been in used over a long period of time. The chronological extremes in the material consist of middle Mesolithic microliths and a fragment from late Neolithic shaft holes axe, which dates the material to a period between 7000-1800 BC. There is nothing to say that the site has been continuously used during this long period. It is possible to date the bulk of the material to the period between late Mesolithic and early Neolithic.

The site has been partly excavated on two occasions. In 1992 the Kalmar County Museum investigated a minor part of the site. During the investigation about 1700 pieces of lithic material was collected, including some 20 transverse arrow heads, micro blades and micro blades cores. A burnt edge fragment from a hollow edged flint chisel shows activity during the middle Neolithic. From the excavation two 14C samples were dated to the late Mesolithic (Källström 1993).

Not far from the 1992 survey an excavation was conducted in 1998 with students from University of Kalmar. During the investigation some 30 test pits were excavated in the slope towards Hagbyån. During the excavation a thick black layer was unveiled, as most 40 cm thick. The layer contained large quantities of burnt stones and chipped kristianstads flint, fragments from polished flint axes, burnt bones, transverse arrow heads, and one sherd from a funnel beaker (Gurstad-Nilsson 2001). Although the layer was thick, it was impossible to distinguish any stratigraphic subdivision. Therefore, the remains from this layer were interpreted as traces from a specialized single activity or a short settlement phase.

The material from these two surveys showed that the sites function and significance varied over time. The materials also suggest that the large settlement at Hagbytorp in fact consists of a series of smaller activity areas. The area contains sub-areas where the material mainly dates to the late Mesolithic while other parts dates to the early Neolithic. An aggregation of various activities from different times has resulted in a merger into one large habitation area.

At the Hagbytorps site there is a frequent occurrence of

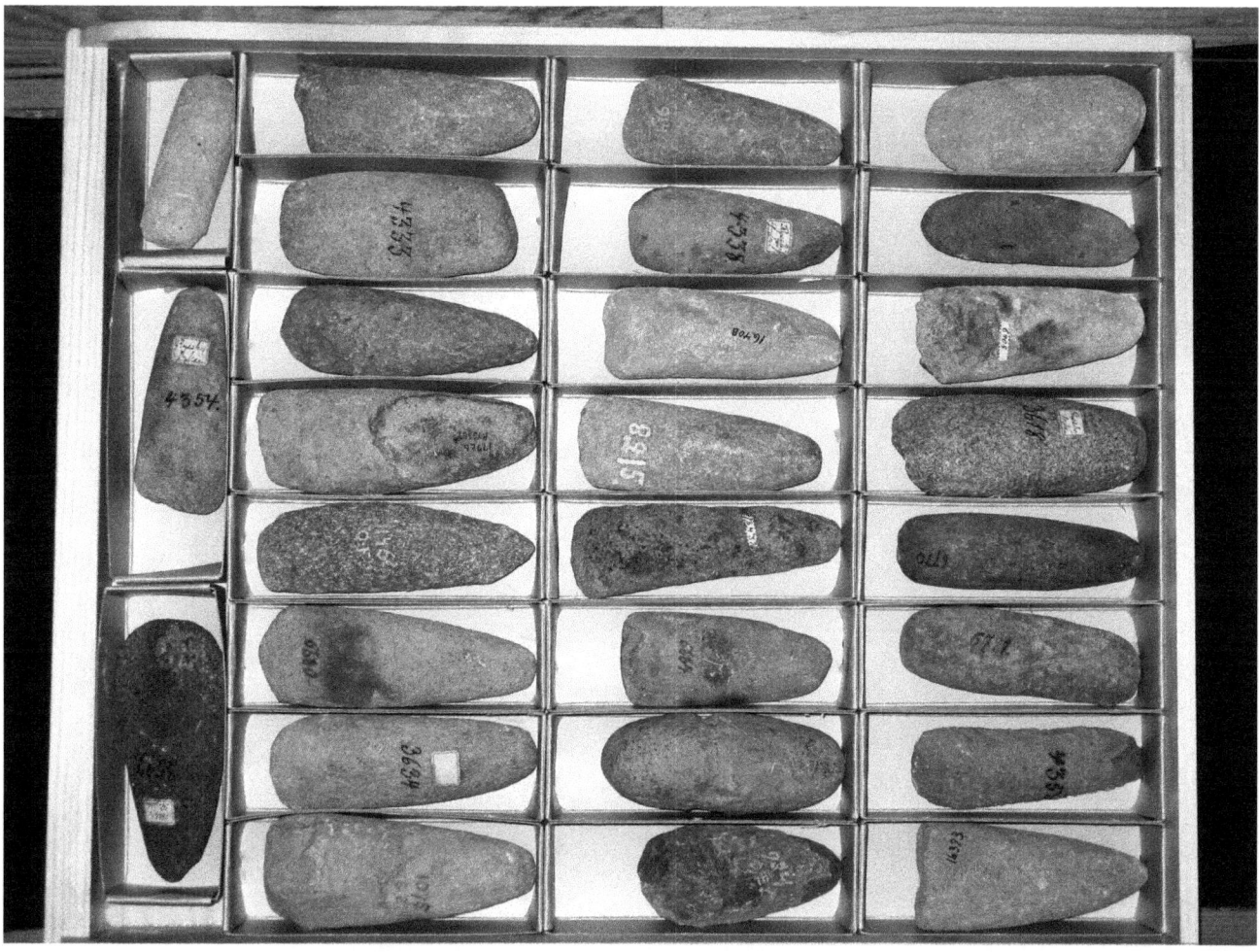

Fig 4 Late Mesolithic pecked stone axes.

chipped flakes from polished flint axes but an absence of whole axes. Instead, a concentration of thin butted flint axes is found on smaller sites in the vicinity.

The settlement pattern during the late Mesolithic into the early Neolithic around the mouths of rivers can be viewed from various perspectives. A traditional interpretation is that people choose to settle in those areas due to the concentration of economic resources from different natural environments. One can also see the mouths of rivers as important from a strategic communications aspect (social, economic, etc.). At the mouths of rivers, there is an opportunity for communications both inland and along the coast. In prehistoric times, the mouths of rivers often come to serve as important nodes in the network through which they linked up with other regions. It is likely during this time that already existing regional networks got a more solidified (Gurstad-Nilsson 2001). These large settlement sites seem to be key points in the landscape during the late Mesolithic and early Neolithic; meeting places for peoples from both the local societies and peoples from more distant areas. On Öland at the opposite side of Kalmarsund the only megalithic tombs in the region were erected, one dolmen in Resmo and three passage graves in Mysinge (Alquist 1822;

Gurstad-Nilsson 2001). It is likely that the establishment of these graves is a result of contacts and networks emerging around the large coastal sites on the mainland. These sites works as gathering sites and act as important sites for social interaction and spreading of new ideas and mentalities among the local societies.

References

Åkerlund, A. 1996. Human responses to shore displacement: living by the sea in Eastern Middle Sweden during the Stone Age, *Arkeologiska undersökningar* 16, Riksantikvarieämbetet.

Alexandersson, K. 2001. Möre i centrum. Mesolitikum i sydöstra Kalmar län. I Möre historien om ett småland. In Magnusson G. (ed.) *E22-projektet Kalmar läns museum*, 111-128.

Alexandersson, K. 2007. Why use different raw materials? Raw material use during the Late Mesolithic to middle Neolithic along the coast of Kalmarsund. In Larsson M. & Pearson M.P. (eds.), *From Stonehenge to the Baltic: Living with cultural diversity in the third millennium BC*. 1692, 35-39. Bar international Series.

Ahlqvist, A. 1822. Ölands Historia och Beskrifning. Första bandet. Faksimilutgåva 1979.

Gurstad-nilsson, H. 2001. En neolitisering – två förlopp. Tankar kring jordbrukskulturens etablering i Kalmarsundsområdet. In Magnusson G. (ed.), *E22-projektet Kalmar läns museum*, 129-164. Kalmar läns museum, Kalmar.

Källström, M. 1993. Hagbytorp, en basboplats från jägarstenålder. Kalmar län, årsbok för kulturhistoria och hembygd. Årgång 78, 36-41. Kalmar, Kalmar läns museum.

Karsten, P. Att kasta yxan i sjön: En studie över rituell tradition och förändring utifrån skånska neolitiska offerfynd. *Acta archaeologica lundensia series* 8, 23. Almquist and Wiksell international, Lund

Persson, P. 1999. *Neolitikums början: Undersökningar kring jordbrukets introduktion i Nordeuropa*. Diss. Kust till kust böcker nr 1, serie B nr. 11GOTARC, Göteborg

Svensson, N-O. 2001. Strandlinjer och strandförskjutning i Möre. In Magnusson G. (ed.) *I Möre historien om ett småland,* 73-110. E22-projektet, Kalmar, Kalmar läns museum.

CHAPTER 11
WALKING WITH COWS:
HUMAN-ANIMAL RELATIONSHIPS IN EARLY NEOLITHIC BRITAIN

Lara Bishop

University of Manchester

Abstract: *Investigation of the relationship between humans and animals in Neolithic Britain is a growing area of research with the potential to greatly enhance our understanding of society at this time. By examining the deposition of animal remains at Early Neolithic mortuary sites in southern Britain some interesting aspects of depositional practices were revealed. A new contextual analysis of faunal deposition at two types of sites - stone chambered tombs and earthen long barrows – has highlighted several important features: the species deposited; their depositional contexts; and their associations with other types of deposit. A key aspect of depositional practices in the Early Neolithic is the prevalence of cattle. Their frequency and prominence suggests they were of great importance to people, possibly far more closely integrated into daily life than any other species due to the nature of subsistence practices and the value of cattle as a symbolic resource, and this may be reflected in their treatment at these sites. It is also interesting that there are a variety of species present, including what we classify as wild and domestic species; but how did people in the Neolithic view the animals they chose to place at these sites? By exploring the possible symbolic associations these animals may have held and re-evaluating our definition of 'domestic' and its usefulness as a concept we can begin to suggest how people ordered the world around them.*

Keywords: *human-animal interaction, early Neolithic, cattle, stone chambered tombs, earthen long barrows*

In the earlier Neolithic, animal remains often form a significant aspect of the assemblage of material from 'mortuary' sites, namely stone chambered tombs and earthen long barrows. These sites have dominated the study of the Neolithic in Britain, no doubt due in part to their visibility in comparison with other types of archaeological deposit from this period and the sharp contrast between these and preceding Mesolithic sites. With the focus of much of the research into Early Neolithic life in Britain centred on mortuary sites and the deposits within, it is unsurprising that the principle interpretive approaches to the period have been based on this evidence (e.g. Barrett 1989; Bradley 1984; Jones 2005; King 2003; Shanks and Tilley 1984; Thomas 2000). The majority of these interpretations have drawn heavily on the presence of human skeletal material and have tended to overlook the animal remains recorded. It is likely that this focus has masked the significance of animals as both a practical and symbolic resource for people in the Early Neolithic. Having recently re-assessed the faunal reports from thirty-three chambered tombs and earthen long barrow sites in southern Britain, it is clear that animals frequently formed a significant aspect of the depositional activity. Indeed, they often occurred in greater amounts than human remains and sometimes animals were deposited at long barrows where no human remains were present.

The earliest studies that considered animals in Early Neolithic Britain tended to concentrate on determining the morphological features of the domestic and wild versions of species and viewed them purely as a record of subsistence practices (e.g. Jewell 1963; Legge 1981).

Interpretations of animal remains in connection with the human dead tended to revolve around the idea of food offerings, provisions for the afterlife, sacrifices or feasting (King 2003, 131). However, if animals were deposited as grave goods, activity would be expected to focus on human remains (Thomas 1991, 34). The similarities between the treatment of human and animal remains at chambered tombs and earthen long mounds suggests this may be far too simplistic an explanation, particularly as this was a period in which the introduction of new species of animal and new kinds of material culture must have altered the way in which humans perceived the world around them. In recent years investigations of the relationships between humans and animals in the Neolithic have become far more central to debates about the period (e.g. Serjeantson and Field 2006). Indeed, the situation has changed markedly in the past decade with the focus shifting to developing an understanding of the meanings of animals within the archaeological record (e.g. Field 2006; Jones and Richards 2003; Pollard 2006; Ray and Thomas 2003).

The growing interest in studies of human-animal relationships within archaeology has drawn on developments in anthropological research, in which studies of this kind have been a focus for some time (e.g. Evans-Pritchard 1940; Ingold 1984; 1988; Morris 2000; Willis 1990). This work has demonstrated not only the variety of relationships that are possible, and the importance of these connections for the people involved, but also highlighted that our modern western view of animals, and nature more generally, is itself socially and historically contingent. An excellent illustration of this is the case of the Uduk of the

Sudanese-Ethiopian border area in east Africa who see themselves as firmly situated within the wider animal world (James 1990). They classify themselves as part of the 'great family of hoofed creatures and kin to the wild antelope', demonstrating that they have a completely different system of classification to that developed by western science (James 1990, 198). Interestingly, in their origin myth they believe a dog brought them fire, taught them language and gave them spears for defence (James 1990, 200), suggesting that they do not view humans as the preeminent species. The example of the Hopi Indians in south-west America is also of note because they believe that the spirit world and this world co-exist in the same space; the spirit world is described as the essence of this world (Bahti 1990, 135). Their relationships with animals are primarily conditioned by that animal's role in the spirit world rather than by day to day interaction in this world, and those animals that are significant in the spirit world are always treated with respect in this one regardless of whether they have what we might call a 'practical' purpose (Bahti 1990, 138-139). This is particularly interesting because the animals that would be represented in the archaeological record would not necessarily be connected with daily subsistence and this has important implications for understanding the animals deposited at Early Neolithic mortuary sites. Crucially, these brief examples illustrate is that our modern western pre-conceptions about the roles of animals, and in particular their subservient status, may be inappropriate if applied to the Neolithic when society may not have conceived of a divide between humans and animals in the same way we do. Fundamentally, both these studies demonstrate the way in which human lives are interlinked with the lives of the animals around them by a variety of factors, of which subsistence need not be the most important.

Mortuary sites in early Neolithic Britain are usually interpreted as places associated with the veneration of a collective group of ancestors rather than as funerary monuments to specific individuals, based on the collective nature of the burial, the disarticulation of the bones and the ability to re-enter the structure and remove or rearrange the deposits according to various classifications (e.g. Shanks and Tilley 1984; Barrett 1989; Bradley 1998). However, this hypothesis has been questioned, due to disparities between modern ethnographic examples and the archaeological evidence (Thomas 2000; Whitley 2002; Whittle 2003). Alternative interpretations have suggested that these sites acted as places for the transformation of the dead to enable their circulation throughout the landscape, that they were significant as a liminal setting rather than for an ancestral connection (Thomas 2000, 662). It has also been argued that it is still possible to consider these sites as monuments associated with the veneration of ancestors because there is the potential for great variability in the interaction with forebears (Whittle 2003, 128). There is a very real probability that modern analogies do not account for all the possibilities. Therefore, while it is undoubtedly important to consider explanations for the presence of animals at mortuary sites that are not necessarily dependent

on connections with human ancestors, we do not have to completely discount them.

Animals at Chambered Tombs and Earthen Long Barrows

This recent re-assessment of faunal reports from thirty three chambered tombs and earthen long barrows revealed some interesting features of the depositional practices at these sites (Bishop 2008). Figure 1 indicates the percentage of the sites where each species was present and there are several key points to highlight. Firstly, cattle were frequently deposited, occurring at all but one of the sites considered and often deposited in prominent positions. At Beckhampton Road earthen long barrow in Wiltshire three cattle skulls were placed on the old land surface along the central axis of the mound without the presence of human remains at the site (Ashbee *et al.* 1979, 247) and a cattle skull and lower leg bones appeared to have been placed on top of the mortuary house at Fussell's Lodge earthen long barrow, also in Wiltshire (Grigson 1966, 65). Secondly, there is an interesting range of species present at the two types of site; a greater range of species is present at chambered tombs but wild species occurred at a greater number of earthen long barrows, which is particularly noticeable in the case of red and roe deer. Thirdly, the non-native or domestic species account for a high percentage of the faunal remains from all the sites. When the percentage of the assemblage for each site was calculated it indicated that at chambered tombs domestic animals accounted for an average of 83% of faunal material and at earthen long barrows they accounted for an average of 79% of assemblages.

Overall, there was no clear emphasis on the deposition of a particular body part. Skulls or skull fragments occurred at a slightly greater number of sites than other areas of the body but in terms of the amount of skeletal elements teeth and antler occurred most frequently. The prevalence of teeth may be due to preservation, although there are definitely some cases in which teeth seem to have been deliberately incorporated in the blocking material of forecourts and chamber entrances at chambered tombs, such as at Hazelton North, Nympsfield and Belas Knap (Thomas and McFadyen 2010, 108). The use of antler tools in the construction of these sites is likely to have affected the frequency of their deposition, although this does not prevent them being deliberately placed, such as at Beckhampton Road where piles of antler tools were evidently arranged (Ashbee *et al.* 1979, 247-248).

There was a broadly equivalent treatment of human and animal remains in the sense that the majority of skeletal elements were disarticulated and from mature individuals, the notable exception being the remains of juvenile domestic animals, which tended to be partially or fully articulated. For example, there was an almost complete (probably originally complete) peri-natal sheep/goat from the south chamber at Hazelton North in Gloucestershire (Levitan 1990, 209-211), a partial burial of a young pig

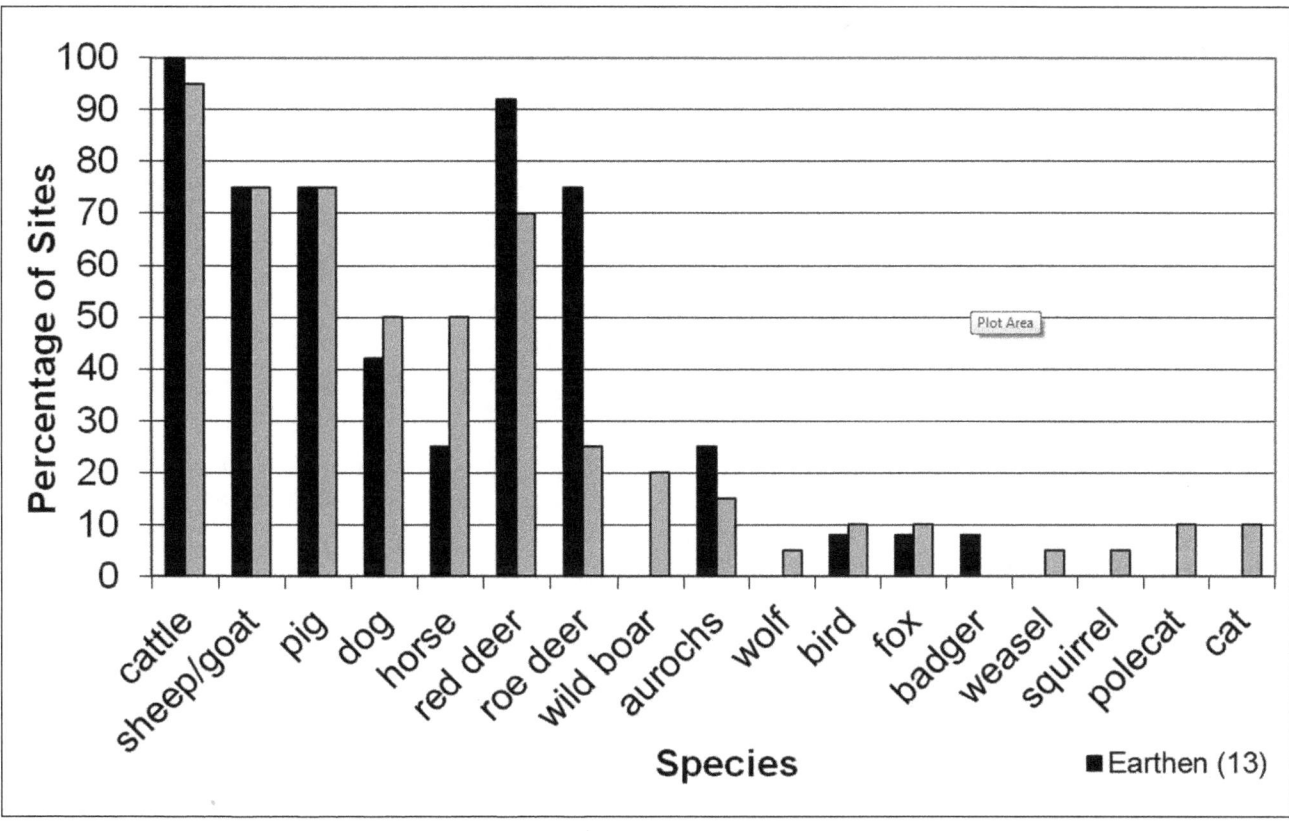

Figure 1: Species occurring at chambered and earthen long barrows, indicating presence of the species at the site rather than absolute numbers of animals or bones

in a pre-barrow pit at Ascott-under-Wychwood chambered tomb in Oxfordshire (Mulville and Grigson 2007, 239) and a partial juvenile dog was recovered from the chamber at West Tump in Gloucestershire (Brickley and Thomas 2004, 5). Interestingly there was no evidence of the deposition of juvenile wild animals, possibly due to the greater difficulty of gaining access to them. The juxtaposition of human and animal remains is also intriguing; there are earthen long barrows where no human remains were deposited (e.g. Thickthorn Down, Beckhampton Road, South Street), but at the chambered and earthen sites where human remains are present there is sometimes equivalence in treatment. At West Tump the juvenile dog was placed next to an adolescent human, showing an association in age and articulation (Brickley and Thomas 2004, 6). In some cases animals seem to be more prominent than humans, as indicated by the cattle skulls and antlers at Beckhampton Road, and therefore they were presumably more focal to the activity at these specific sites. A key point to note is that there were more partially or fully articulated animal remains at chambered tombs and animals were more often found associated with humans at those sites. In contrast, at earthen long barrows human remains were not always present but the amount of animal remains deposited was more consistent throughout the use of sites, indicating that animals were a more constant element in the depositional practices in these locations. There was, however, a greater variation in the composition of the faunal assemblages at earthen long barrows than at chambered tombs, with varying percentages of wild and domestic animals deposited.

It will be useful to consider this differentiation between wild and domestic in more detail because it is clear that wild animals could be deposited in significant locations as well as domestic ones; an aurochs skull was placed on the base of the ditch and vertebra deposited on the old land surface at Thickthorn Down earthen long barrow in Dorset (Drew and Piggott 1936, 82, 93). The activity at the sites considered was broken down into phases of use to aid comparison between sites. Figures 2 and 3 demonstrate that overall more wild animals were deposited at earthen long barrows than chambered tombs and there was a greater variation in the composition of the deposits throughout the activity at earthen sites. This is an obvious contrast with chambered tombs where the majority of the faunal assemblage was comprised of domestic species and this remained quite constant throughout the use of sites. This indicates that not only was there a greater physical connection between humans and animals at chambered tombs, but that domestic species, which were presumably more closely entwined with humans in daily life, were prominent throughout the entire use of the sites. This key difference in the treatment of animals at these two types of site may indicate a difference in their purpose; the prominence of domestic species that were reliant on and intimately connected with humans at chambered tombs, when considered alongside the regularity with which human remains were deposited, suggests that these sites had more of a defined mortuary use. They were evidently connected with the human dead and with the animals that would have been most familiar to those people. Whereas, at earthen long barrows a range of both wild and domestic animals were often the dominant deposit, suggesting that we need to explore the possible meanings of these animals to develop a better understanding of the use of these sites.

Wild versus Domestic?

At this point it is important to consider whether it is appropriate to apply our modern concepts of domestic and wild when we are discussing the early Neolithic. Certainly it is clear from the incorporation of 'wild' animals in chambered tombs and earthen long barrows that these were not seen as completely alien species that had no role in human society. These are concepts that have proved difficult to define, with definitions of domestication having generally been approached from either a biological or social perspective and there is an ongoing debate concerning the intentionality of the process (Russell 2002, 286-287). Biological definitions tend to focus on the control of breeding, which is seen as an intentional human action (Arbuckle 2005, 19), or symbiosis, where animals are attracted to humans and tolerated by them, leading to unintentional domestication (Russell 2002, 287-289). Social definitions tend to focus on the idea of animals as human property (Russell 2002, 290-291). It has been suggested that there were forms of subsistence practice in prehistory unlike any historic or modern analogies and the process of domestication may have taken many different forms (Higgs and Jarman 1972; Redding 2005). Additionally, there are a range of intermediate relationships that cannot be neatly categorised as interaction with either 'wild' or 'domestic' animals (Vigne *et al.* 2005, 3).

The development of the idea of human-animal relationships formed an attempt to move beyond this wild/domestic dichotomy and think about human relationships with non-humans in terms of differing degrees of control over animal lives within a continuum of interaction (Higgs and Jarman 1972, 12). This approach moved away from the idea of domestication as an event that could be detected by morphological change and viewed it as a continually evolving process intentionally created by humans due to pressures of population and environment (Higgs and Jarman 1972, 12). Building on this, it has been contended that in order to discuss the idea of domestication we cannot rely on definitions that are restricted to either biological or social explanations, which tend to mirror the nature/culture dichotomy, but we benefit from approaching the concept of domestication in terms of both nature and culture (Russell 2002, 286). This is due to the difficulty of defining human-animal relationships as wild or domestic and the nature of the domestication process itself, which contains biological and social components (Russell 2002, 286). Based on the idea of human-animal relationships Russell suggests that we should view these as a spectrum of activities that can include domestication, thereby maintaining its importance without devaluing other types of human-animal relationship (Russell 2002, 295). The creation of 'domestic' animals and plants allowed the creation of the 'wild' as a concept and in

particular the domestication of animals allowed particularly powerful metaphors to be created, which had implications not only for human-animal relationships but could also act as an example of new types of human-human interaction (Russell 2002, 296-7). The key point that is relevant to this study is the idea of a spectrum of interaction; domestication can still be seen as important, it is a radically different way of interacting with animals than in, for example, a purely hunting society, but it acknowledges that there are many other forms of relationship. In Neolithic Britain the evidence suggests that 'domestic' animals were important but that a range of other species were also crucial aspects of depositional practice at mortuary sites.

It has been suggested that in the Neolithic wild species do not feature as prominently as domestic ones within Early Neolithic deposits because they were more distant from social practice and therefore did not have the same web of associated meaning as domestic animals (Pollard 2006, 138). In addition, the exchange networks implied by the introduction of domesticated species probably made those animals 'more intriguing and valuable' (Pollard 2006, 139). It is possible that there was no conceptual division between wild and domestic animals and human understanding of animals was based on lived experience rather than abstract conceptual schemes (Pollard 2006, 145); it seems logical that the distance from human social practices affected the frequency with which animals occurred in depositional practices.

If we think about this in relation to cattle remains at mortuary sites; the frequency with which cattle were deposited, the large amount of bones in comparison to other animals, and the prominent locations in which they were often placed, it suggests that they were very closely integrated into human society. This is also supported by the evidence from lipid analysis of pottery which indicates that cattle dairy products were contained within pottery vessels (Copley *et al.* 2005). Dairying would require a very close interaction between humans and cattle and may have contributed to their prominence in depositional activity. It has also been suggested that this is due to the idea of a type of kinship between humans and cattle; that cattle may have been a form of wealth that was used to create alliances, could be used in gift exchange or could even be raided (Ray and Thomas 2003, 41; Russell 1998, 50-51). These transactions would have created a genealogy where the composition of cattle herds could have been viewed as a parallel community that mirrored human kinship relations (Ray and Thomas 2003, 41). The juxtaposition and interchangeable deposition of human and cattle remains at chambered tombs and earthen long barrows suggests that both could intercede on the living community's behalf with human ancestors (Thomas 2003, 38-40). This suggests a more equal relationship between humans and this particular animal.

If we consider how these suggestions affect the interpretation of the other animal remains present at

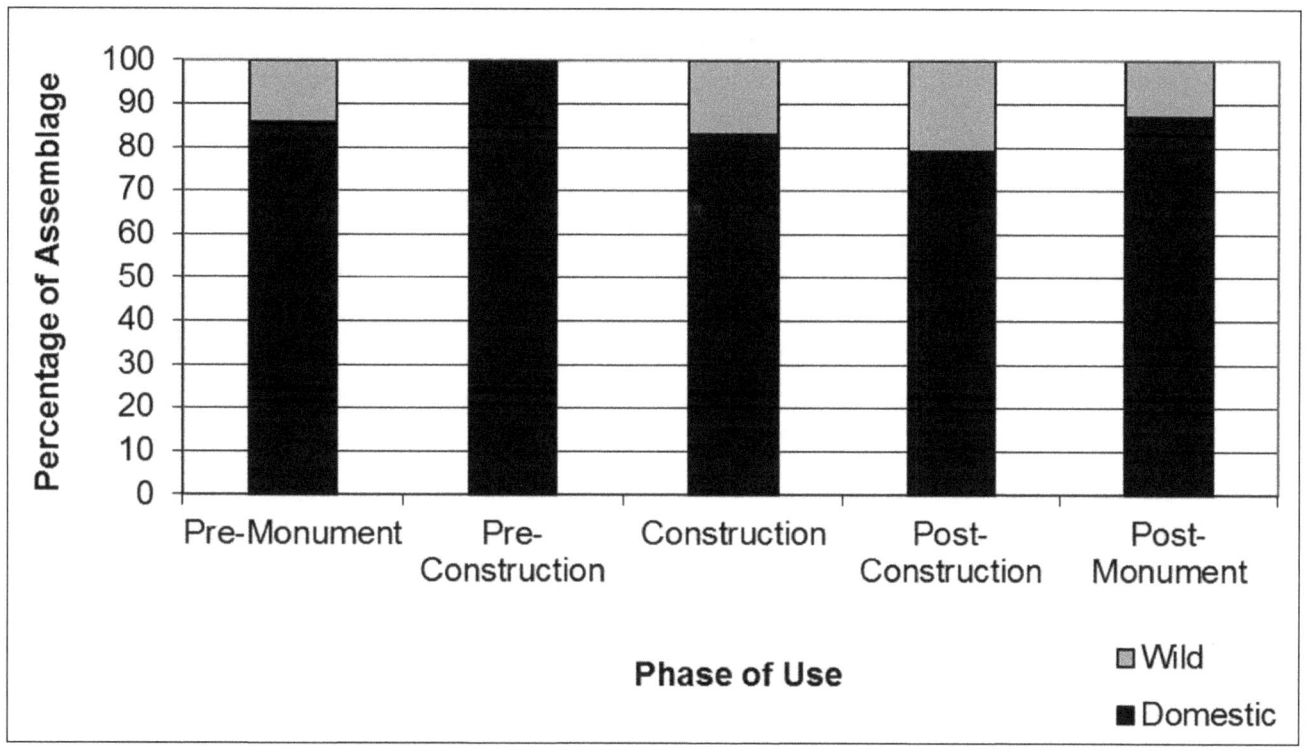

Figure 2: Percentage of wild and domestic animals by phase of use at chambered tombs

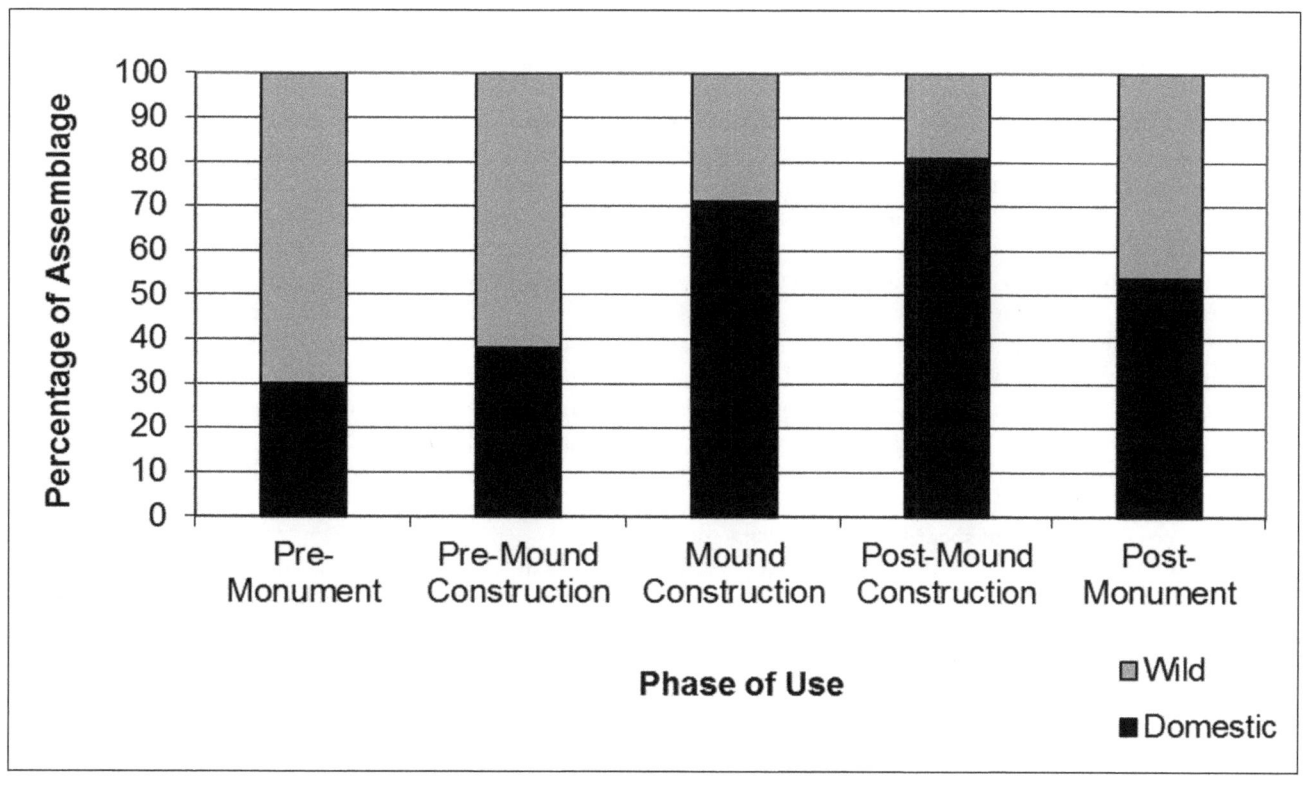

Figure 3: Percentage of wild and domestic animals by phase of use at earthen long barrows

chambered tombs and earthen long barrows perhaps we could think of human relationships with animals in terms of degrees of association. At one end of the scale, cattle were intimately connected with humans on a daily basis, sheep and pig only marginally less so and then deer slightly further removed, yet still important because of the need for antler. Beyond this aurochs, wolves and wild boar were presumably more distantly associated with humans due to their frequency of occurrence, yet appear to have been hunted, and then other animals that rarely appear in the archaeological record, such as bear, were the least associated. This does not assume that humans had no contact with or symbolic values for the most distantly related species, but these were not appropriate values to be incorporated in earthen long barrows or chambered tombs.

The Meaning of Animals

But why was this variety of species deposited at mortuary sites – what did they mean to people in the Neolithic? For some communities, such as in rural Malawi or many circumpolar societies, spirits and sometimes ancestors can take the form of animals (Morris 2000, 221; Ingold 1986, 245). It could be that at least some of the animals deposited at these sites may have represented spirits, indicating that various spiritual agencies could be envisaged within Neolithic society. To turn to another anthropological analogy, it has been observed that totemism and animism can be combined within one society, even though these seem to be contrasting methods of classification (Descola 1996, 87-88). However, this is achieved by distinguishing between two distinct groups of non-human animals, the one understood through totemism and the other through animism (Descola 1996, 87-88). It seems that this way of dividing non-humans may be useful for understanding human-animal relationships in the Early Neolithic. Day-to-day experience may have led to the formation of personal relationships with animals that were closely connected with human society, such as cattle and dogs, and so they were deposited at chambered tombs and earthen long barrows because of this human connection. Other animals were more distantly associated with humans and this element of mystery may have led to the formation of spiritual or cosmological associations; it was this that made them appropriate deposits to be incorporated in earthen long barrows and chambered tombs. These ideas may link into the differences in the range of species deposited at the two types of site and their possible differences of purpose; chambered tombs had a greater focus on the deposition of human remains and the animals that were intimately connected with them, while deposition at earthen long barrows focused on less closely associated species.

Finally, it is important to consider the degree of association between the living animal and the specific element of the animal that was deposited at mortuary sites. To what extent was the element seen to represent the whole animal? Or was the specific part of the animal seen to possess particular properties of its own? In order to address these issues it

seems logical to suggest that it might be possible to look for evidence of alteration to the element of the animal. Whilst a large amount of the animal remains deposited at chambered tombs and earthen long barrows appears to be unaltered, apart from disarticulation, there are several examples of worked bone and many examples of antlers that show signs of wear. A few examples from earthen long barrows include: worn antlers from South Street (Ashbee *et al.* 1979, 268-269); ten antler picks and two rakes from Beckhampton Road (Ashbee *et al.* 1979, 247-248); a small bone chisel from Giant's Hills 1 (Phillips 1935, 58); and a purposefully shaped antler and a possible unfinished antler comb from Thickthorn Down (Drew and Piggott 1936, 87, 93). From chambered tombs: at Hazleton North four bone beads and two bone points (Saville 1990, 178-180); at Sale's Lot three fragments of bone pins (O'Neil 1966, 31-32); at Ascott-Under-Wychwood two antler picks and one worked antler (Mulville and Grigson 2007, 244); at Wayland's Smithy sixteen worn antlers (Whittle 1991, 88); and at Penywrylod an apparent bone flute/pipe type instrument (Britnell and Savory 1984, 27). In these cases we can see there was little difference in the types of worked bone and antler deposited at the two types of site.

Perhaps worked bone and antler by their physical alteration were transformed in their meaning(s) as well. This is easy to imagine with objects such as the possible flute from Penywrylod, which is not so recognisable as bone, but slightly more problematic with antler tools, which still closely resemble the original item. At the other extreme articulated animals could be seen as retaining the qualities or ideas associated with the living animal. The large amount of disarticulated animal bone is interesting; the slightly larger amount of skulls or skull fragments of the most commonly deposited species may be explained because this was the most recognisable part of the animal, it could have been representative of the whole animal and possibly imply the presence of the spirit of the animal (Field 2006, 130). However, the majority of the animal bone deposited did not show a clear emphasis on the deposition of a particular body part, perhaps in part due to the effect of preservation or recording, but quite possibly because this was not a necessary distinction. Therefore, if we follow this line of thought, it could be argued that any element of the animal may have been seen as representative of the whole and be indicative of the meanings associated with it.

It is possible to find some parallels for this discussion in the interpretation of the antler masks from Early Mesolithic Star Carr (Conneller 2004). The traditional interpretation of these masks was that they were either a hunting disguise or a ritual dance mask, but in both cases concealment was thought to be the aim (Conneller 2004). Conneller draws on the work of Viveiros de Castro in his study of Amerindians, who see all humans, many animals and objects as being internally identical (all have an essence or spirit) and the specific properties or attributes of each animal or human are located in the external body; humans can take on the attributes of an animal by wearing an aspect of its body

(Conneller 2004, 43). At Star Carr people used animals for food, tools and clothing, animal bodies were broken down and new assemblages of human-animal bodies were created (Conneller 2004, 47-48). The use of animals in this way could have been seen as a way of extending the human body by taking on the properties from the specific part of the animal used (Conneller 2004, 47-48). Although there is evidently no direct connection between Early Mesolithic Star Carr and Early Neolithic mortuary sites this idea does illustrate that our concepts of the distinction between human and animal bodies may need to be broken down when we are thinking about the Neolithic. Maybe we need to think again about what the physical association between humans and animals means in depositional practice and consider that perhaps new types of beings, neither completely human nor animal, were being enacted by this process.

Conclusion

This paper has suggested that the lived connection between humans and animals may have determined how humans classified animals, with the animals that were closely integrated into human society those that were most frequently deposited at the sites considered in this study. Humans could have conceived of degrees of association with species of animals, ranging from close daily interaction, such as that suggested for cattle, to infrequent encounters. The archaeological record suggests that the closely associated species may have been valued for their connection with human society while the distantly associated species may have held mystical or spiritual properties. Interestingly, when the nature of the deposits at earthen long barrows and chambered tombs were compared they indicated a greater focus on the deposition of closely associated species of animal with human remains at chambered sites, whereas at earthen sites human remains occurred less frequently and there was a greater proportion of distantly related species. This may indicate that there was a difference in the purpose of the two types of site. It appears that there was no particular focus on the deposition of specific skeletal elements, which suggests that any element could represent the animal and its associated meanings. A detailed investigation of animal remains from what are usually considered to be mortuary sites for the processing and disposal of the human dead can provoke new ways of thinking about both humans and animals in the Neolithic.

Bibliography

Arbuckle, B.S. 2005. Experimental Animal Domestication and its Application to the Study of Animal Exploitation in Prehistory. In J.D. Vigne, J. Peters and D. Helmer, (eds.) *The First Steps of Animal Domestication.* New Archaeological Approaches, 18-33. Oxford, Oxbow.

Ashbee, P., Smith, I. and Evans, J.G. 1979. Excavation of three long barrows near Avebury, Wiltshire. *Proceedings of the Prehistoric Society* 45, 207-300.

Bahti, M.T. 1990. Animals in Hopi Duality. In R. Willis (ed.), *Signifying Animals. Human Meaning in the Natural World.* 134-9. London, Unwin Hyman.

Barrett, J. 1989. The living, the dead and the ancestors: Neolithic and Early Bronze Age mortuary practices. In J. Barrett and I. Kinnes (eds.), *The Archaeology of Context in the Neolithic and Bronze Age*, 30-41. Sheffield, University of Sheffield.

Bishop, L.D. 2008. *Lambs to the Slaughter? A Re-assessment of Animal Remains from Early Neolithic Mortuary Sites in Southern Britain.* Unpublished MPhil Thesis, University of Birmingham.

Bradley, R. 1984. *The Social Foundations of Prehistoric Britain*, London, Longman.

Bradley, R. 1998. The Significance of Monuments. London, Routledge.

Brickley, M. and Thomas, R. 2004. The young woman and her baby or the juvenile and their dog: Reinterpreting osteological material from a Neolithic Long Barrow. *Archaeological Journal* 161, 1-10.

Britnell, W.J. and Savory, H.N. 1984. *Gwernvale and Penywyrlod: Two Neolithic Long Cairns in the Black Mountains of Brecknock*, Cambrian Archaeological Monographs No 2, Caambridge, Cambridge Archaeological Association.

Conneller, C J 2004. Becoming Deer: Corporeal Transformations at Starr Carr. *Archaeological Dialogues* 11(1), 37-56.

Copley, M.S., Berstan, R., Mukherjee, A.J., Dudd, S.N., Straker, V., Payne, S. and Evershed, R.P. 2005. Dairying in Antiquity. III. Evidence from Absorbed Lipid Residues Dating to the British Neolithic, *Journal of Archaeological Science* 32, 523-46.

Drew, C.D. and Piggott, S. 1936. The Excavation of Long Barrow 163a on Thickthorn Down, Dorset, *Proceedings of the Prehistoric Society* 2, 77-96.

Descola, P. 1996. Constructing Natures. Symbolic Ecology and Social Practice. In P. Descola and G. Pálsson, (eds.) *Nature and Society. Anthropological Perspectives*, 82-102. London, Routledge.

Evans-Pritchard, E.E. 1940. *The Nuer. A Description of the Modes of Livelihood and Political Institutions of a Nilotic People.* Oxford, Oxford University Press.

Field, D. 2006. Earthen Long Barrows. The Earliest Monuments in the British Isles, Stroud

Grigson, C 1966. The Animal Remains from Fussell's Lodge Long Barrow. In P. Ashbee, (ed.), The Fussel's Lodge Long Barrow, 1957, *Archaeologia* 100, 63-73.

Higgs, E.S. and Jarman, M.R. 1972. The Origins of Animal and Plant Husbandry. In E.S. Higgs (ed.) *Papers in Economic Prehistory. Studies by Members and Associates of the British Academy Major Research Project in the Early History of Agriculture.* 3-13 Cambridge, Cambridge University Press.

Ingold, T. 1984. Time, Social Relationships and the Exploitation of Animals: Anthropological Reflections on Prehistory. In J. Clutton-Brock and C. Grigson (eds.) *Animals and Archaeology: 3. Early Herders and Their*

Flocks, British Archaeological Reports International Series 202, 3-12.

Ingold, T. 1986. *The Appropriation of Nature*. Manchester, Manchester University Press.

Ingold, T 1988. The Animal in the Study of Humanity. In T. Ingold (ed.) *What is an Animal?*. 84-99 London, Unwin Hyman.

James, W 1990. Antelope as Self-Image among the Uduk. In R. Willis (ed.) *Signifying Animals. Human Meaning in the Natural World*. 196-203, London, Routledge.

Jewell, P 1963. Cattle from British Archaeological Sites. In A.E. Mourant and F.E. Zeuner, (eds.) *Man and Cattle*. 80-101, London, Proceedings of a Symposium on Domestication at the Royal Anthropological Institute 24-26 May 1960.

Jones, A. 2005. Lives in Fragments? Personhood and the European Neolithic, *Journal of Social Archaeology* 5(2), 193-224.

Jones, A. and Richards, C. 2003. Animals into Ancestors: Domestication, Food and Identity in Late Neolithic Orkney. In M. Parker Pearson (ed.) *Food, Culture and Identity in the Neolithic and Early Bronze Age*, British Archaeological Reports International Series 1117, 45-51.

King, M. P. 2003. *Unparalleled Behaviour: Britain and Ireland during the 'Mesolithic' and 'Neolithic'*. British Archaeological Reports British Series 355.

Legge, A.J. 1981. Aspects of Cattle Husbandry. In Mercer, R. (ed.) *Farming Practice in British Prehistory*. 169-181 Edinburgh, Edinburgh University Press.

Levitan, B. 1990. The Non-Human Vertebrate Remains. In A. Saville (ed.) *Hazelton North, Gloucestershire, 1979-82: The Excavation of a Neolithic Long Cairn of the Cotswold-Severn Group*. 199-213, London, English Heritage.

Morris, B. 2000. *Animals and Ancestors. An Ethnography*. Oxford, Berg Morris.

Mulville, J. and Grigson, C. 2007. The Animal Bones. In D. Benson and A. Whittle (eds.) *Building Memories. The Neolithic Cotswold Long Barrow at Ascott-Under-Wychwood, Oxfordshire*. 237-53, Oxford, Oxbow.

O'Neil, H. 1966. Sale's Lot Long Barrow Withington, Gloucestershire 1962-65. *Transactions of the Bristol and Gloucestershire Archaeological Society* 85, 5-35.

Phillips, C.W. 1935. The Excavation of the Giants' Hills Long Barrow, Skendleby, Lincolnshire. *Archaeologia* 85, 37-106.

Pollard, J. 2006. A Community of Beings: Animals and People in the Neolithic of Southern Britain. In D. Serjeantson and D. Field (eds.) *Animals in the Neolithic of Britain and Europe*. 135-48 Oxford, Oxbow.

Ray, K. and Thomas, J. 2003. In the Kinship of Cows: the social centrality of cattle in the earlier Neolithic of Southern Britain. In M. Parker Pearson (ed.) *Food,*

Culture and Identity in the Neolithic and Early Bronze Age, British Archaeological Reports International Series 1117, 37-44.

Redding, R.W. 2005. Breaking the Mold: a Consideration of Variation in the Evolution of Animal Domestication. In J.D. Vigne, J. Peters and D. Helmer (eds.), *The First Steps of Animal Domestication. New Archaeological Approaches.* 41-48 Oxford, Oxbow.

Russell, N. 1998. Cattle as Wealth in Neolithic Europe: Where's the Beef? In D. Bailey (ed.) *The Archaeology of Value. Essays on Prestige and the Processes of Valuation,* British Archaeological Reports International Series 730, Oxford, 42-54.

Russell, N. 2002. The Wild Side of Animal Domestication. *Society and Animals* 10 (3), 285-302.

Saville, A. 1990. *Hazelton North, Gloucestershire, 1979-82: The Excavation of a Neolithic Long Cairn of the Cotswold-Severn Group*. London, English Heritage.

Serjeantson, D. and Field, D. (eds.) 2006. *Animals in the Neolithic of Britain and Europe*. Oxford, Neolithic Study Group.

Shanks, M. and Tilley, C. 1984. Ideology, Symbolic Power and Ritual Communication: a reinterpretation of Neolithic mortuary practices. In I. Hodder (ed), *Symbolic and Structural Archaeology*. 129-154 Cambridge, Cambridge University Press.

Thomas, J. 1991. Reading the Body. In P. Garwood, D. Jennings, R. Skeates and J. Toms (eds.), *Sacred and Profane. Proceedings of a Conference on Archaeology, Ritual and Religion*. Oxford 1989, Oxford, 33-42.

Thomas, J. 2000. Death, Identity and the Body in Neolithic Britain. *Journal of the Royal Anthropological Institute* 6(4), 653-668.

Thomas, R. and McFadyen, L. 2010. Animals and Cotswold-Severn Long Barrows: a Re-examination. *Proceedings of the Prehistoric Society* 76, 95-113.

Vigne, J.D., Helmer, D. and Peters, J. 2005. New Archaeolzoological Approaches to Trace the First Steps of Animal Domestication: General Presentation, Reflections and Proposals. In J.D. Vigne, J. Peters and D. Helmer (eds.), *The First Steps of Animal Domestication, New Archaeological Approaches*. 1-16 Oxford, Oxbow.

Whitley, J. 2002. Too Many Ancestors. *Antiquity* 76, 119-26.

Whittle, A. 1991. Wayland's Smithy, Oxfordshire: Excavations at the Neolithic Tomb in 1962-63 by R.J.C. Atkinson and S. Piggott. *Proceedings of the Prehistoric Society* 57(2), 61-101.

Whittle, A. 2003. *The Archaeology of People. Dimensions of Neolithic Life*. London, Routledge.

Willis, R. (ed.) 1990. *Signifying Animals. Human Meaning in the Natural World*. London, Unwin Hyman.

CHAPTER 12
THROUGH THE LOOKING GLASS, MICROSCOPY TO EXAMINE THE LARGER PICTURE:
NEOLITHIC LITHICS FROM BRITAIN

Jolene Debert

Mount Royal University

Abstract: *Within the field of lithic microwear studies many authors have redesigned methodologies, applied new techniques, run blind tests, and conducted experimental work (Debert and Sherriff 2007; Debert 2010; Donahue and Burroni 2004; Evans and Donahue 2005; Hogberg et al. 2009; Lombard 2005; Newcomer et al. 1987; Odell 1988; Rots and Williamson 2004; Shea and Klenck 1993; Ungerhamilton 1989; Wadley et al. 2004; Wadley and Lombard 2007; West and Louys 2007); however, very few have looked beyond the microscopic to the larger picture. The validity of the above types of papers is not in question, what is, are author's reservations about taking the next interpretative step.*

This paper has two main objectives. Firstly, this paper seeks to examine this issue and bring it to the forefront of microwear studies. The second is to provide an example of the type of interpretation that is possible with microwear results. A collection of lithics from four early Neolithic timber structures from Britain will serve as the sample. These are the structures of Parc Bryn Cegin and Llandegai (in Wales), White Horse Stone and Yarnton (in England). This collection was chosen as the usewear results provide tantalising clues as to the role of the timber structures in early Neolithic Britain. As the architecture provides the stage, the finds provide the means of extracting the activities and motives of the distant actors.

Keywords: *Lithic analysis, Early Neolithic, Britain, Timber structures, Usewear, Microwear, Stone tools*

Introduction to the Issue

Microwear analysis originates with the Russian researcher S.A. Semenov; the first person to use magnification to examine the surface of flint. With the translation of his Prehistoric Technology into English, in 1964, a new field of research was born (Semenov, 1964). After initial optimism, microwear analysis suffered a crisis of faith, similar to other scientific techniques used in archaeology at the end of the processual paradigm. It is only now with renewed vigour, addition of new techniques, acceptance of blind tests and countless experimental studies that lithic microwear has regained its position in scientific archaeological research (Andrefsky, 1998, Newcomer et al., 1987, Cook, 1987, Unrath et al., 1986, Vaughan, 1985, Odell and Odell-Vereecken, 1980).

However, this acceptance has not translated to widespread use. Lithic analysis may no long be relegated to the appendixes of excavation reports but lithic microwear remains a specialism of a specialty. The majority of microwear papers discuss techniques, set out to prove the validity of a methodology or describe the analysis of a single set of lithics. Very few microwear studies try to address any questions larger then the collection they analyse. It is therefore our own failing as microwear specialists that have restricted our impact and readership.

The majority of lithisists will receive a collection from a site they probably never excavated any maybe only visited, finish their examination, produce a report for the excavating archaeologist and possibly publish their methodology or narrow findings in a subject specific journal. Thus the lithisists is not involved in the larger questions. Interpretations about the nature or use of the site, its meaning in relation to the wider landscape and community are left to other archaeologists. I am careful to note that there are a number of exceptions to this stereotype; however, these impressive interpretive narratives are not standard in the lithic community. So why is it then that the majority of lithisists and more specifically usewear specialists are content to leave the interpretation to another? This is a question I will leave the reader to ponder.

The objective of this paper is not just to highlight the issue but to serve as an example of the integration of lithic microwear studies in an interpretive narrative. Here, functional usewear will be woven into the interpretation of the early Neolithic timber structures and their Neolithic context. As such there will be no attempt to discuss methodology or justify its use. Instead the reader will simply be directed to the already large body of work doing just that. Like other microwear papers only a finite number of flints were actually examined. The flints examined are from the timber structures Parc Bryn Cegin and Llandegai in Wales and Yarnton and White Horse Stone in England.

Introduction to Timber Structures

Several large timber structures are now known from the early Neolithic in Britain. This is due largely to the increase in the number of salvage excavations and to changes in excavation techniques, which are now exposing larger areas. The sudden appearance of these large timber structures has sparked considerable debate about the nature of the early Neolithic in Britain (Tipping et al., 2009, Sheridan, 2010, Thomas, 2007, Sheridan, 2007, Darvill, 2007).

These structures date to the initial Neolithic, between 4000-3650 BC (Barclay et al., 2002a, Brophy, 2007, Debert, 2010, Garton, 1991, Hayden and Stafford, 2006, Hey and Bell, 2000, Kenney and Davidson, 2006, Lynch and Musson, 2004, Ralston, 1982). They are post built, sometimes employing bedding trenches. They all appear to have been burnt down at the end of their use. The structures are rectangular, some with rounded ends, with lengths exceeding 10 metres and widths often over 5 metres (Barclay et al., 2002a, Darvill, 1996, Fairweather and Ralston, 1993, Fraser, 2006, Kenney and Davidson, 2006, Lynch and Musson, 2004, Ralston, 1982). The division of space within each structure is quite similar, with the larger examples having five discrete areas.

Parc Bryn Cegin and Llandegai are two of the smallest timber structures with only three internal areas. These structures are located within 500 metres of each other. White Horse Stone in Kent is also paired with another timber structure; the poorly preserved structure of Pilgrim's Way, 240 metres away. Pilgrim's Way is not included in this discussion as no utilized flints were recovered during its excavation (Debert, 2010). Finally, Yarnton appears to have been isolated with the only possible contemporary structure being that of a smaller round structure to the southwest.

Microwear Examination

As mentioned above, the purpose of this paper is not to defend the methodology used in the examination of the flints and therefore it will not be discussed in depth. All flints from the four structures were examined by 10x hands lens. All pieces that showed signs of edge wear, breakage, heat exposure or modification were retained for examination with scanning electron microscopy (SEM). For a detailed description of this methodology please refer to the author's previous paper on the subject (Debert and Sherriff, 2007).

Parc Bryn Cegin

Of the twenty-seven flints recovered from Parc Bryn Cegin six flints were selected for microwear analysis (Kenney, 2006, Lynch and Musson, 2004). The examination revealed four had been utilized. These were: a haftable scraper used to scrape soft woods, dry hide or fibrous plants; two sickle blades; and one spokeshave used to scrape wood or fresh bone into a concave shape.

Llandegai

Five utilized flints were identified out of the twenty-six from Llandegai. These were three scrapers: two used on bone, dry wood or dry antler and one on fresh bone, fresh antler or hard woods; and two knifes used to cut fresh meat or non-fibrous plants.

White Horse Stone

Of the four hundred and ninety flints recovered from White Horse Stone three pieces had identifiable microwear. All three were used for a scraping function, though one was poorly modified and probably only used a short time. The other two flints were a unidirectional scraper used on medium-soft materials (fresh wood, dry hide, dried or frozen meat and fibrous plants) and a multi-edged scraper used on medium-soft (as above) to soft materials (meat, fresh hide, and non-fibrous plants).

Yarnton

Fifteen flints were recovered from Yarnton; two had usewear. Interestingly both flints are whitish flint and come from the same post fill context. One is a broken blade and the other is a microblade. Both flints had the same function. Each tool had two parallel cutting surfaces that ran the length of the blade or microblade. These edges were used for the cutting of medium-soft materials. Given the degree of polish development they were probably used on silica rich plants, in other words, they were likely used as sickle blades.

Weaving a Narrative

As mentioned earlier, this paper is attempting to push beyond the descriptive nature of most microwear studies; as such I will now weave the microwear results within a discussion of the timber structures. As Tim Darvill (1996: 79) has stated, "buildings do not just happen, they are deliberately created within the context of the systems of social action." Likewise lithic tools are a medium of this type of action and can be used to inform about the wider context.

The association of memory with material and/or object has been developed by several authors though most often dealing with stone or metal. It is the experience, place and difficultly encountered in obtaining the material that is transplanted and retained with the object (Thomas, 1999, Bradley and Edmonds, 1993). "The object itself is but an object, it is ultimately the connotations connected to this object that make an object inalienable or even sacred" (Wentink, 2006: 84). I would suggest that the same attachment of place and memory is possible with organic materials. A parallel can be drawn between the retention of rock sources, used in the production of stone axes and the harvesting of timbers used in the construction of timber structures. Thereby the memory of place is transplanted with the timbers to be re-erected with other similarly imbued timbers.

These structures may simply be seen as an amalgamation of timber, thatch, wattle and daub, possibly collected from great distances. But the collection of these resources would have brought people together over vast areas. At the same time, if the timbers retained the essence of their location or the memory of their felling and transport then, the timber structure is their world concentrated in one transmorphic place. In other words, the erection of a timber structure is a physical manifestation of the building of a new Neolithic community. Being timber the posts making up the building could easily have been elaborated with carvings or embellishments to help retain this memory. In fact the scrapers at White Horse Stone, which show used on fresh wood, could have been used in the modification of the timbers. Similar wood working scarpers were found at Parc Bryn Cegin and Llandegai. To quantify the validity of this idea; of the fourteen flints in this study with usewear, eight (57%) were used to scrape wood, though one was a concave scraper.

It is thought that these structures stood, with minimal or no reconstruction for nearly a hundred years. As such we cannot look at these structures simply through the act of erection. Instead they became an integral part of the new Neolithic community. They would have been used for generations after the original building. It is quite possible that the malleable nature of the timber making up the structures resulted in their continual modification thus making the structure relevant to subsequent generations. The fact that some of the microwear points to use on dry wood supports the idea of post construction modification.

Continuing on with the ideas that the timber posts have been imbued with symbolic meaning, retaining their sense of place, and their amalgamation in the structure as a microcosm of the wider landscape; I want to consider the visitor's experience. The interior is obscured from outside view in almost every excavated early Neolithic timber structure (notable exception is Lockerbie). This means the visitor cannot see the internal space until they fully enter. Likewise, once the visitor has penetrated the internal space of the structure they can no longer see the outside. In a sense the visitor is transported from one world into another just by the restriction of sight (Bradley, 2003, Barclay et al., 2002a). In other words it was important to either exclude the outside landscape from the constructed space within the structure or to exclude some people from specific activities or events within the structure, or a combination of the two (Nanoglou, 2008, Fletcher, 2008, Hardin, 2004).

Several authors have argued that Barnhouse, Maes Howe and Stenness all represent different manifestations of a cosmological model (Ruggles and Barclay, 2000, Richards, 1996, Richards, 1993). I mention this late Neolithic landscape as the argument relies on the restriction of movement, the emphasis of certain directions and themes

all encountered in the three sites. To discuss architecture as "socially significant and symbolically-loaded" (Barclay et al., 2002b: 124) is now quite accepted. Therefore, the idea that the timbers of the timber structures are embodied with meaning and possibly modified for remembrance is relevant.

Our ability to access movement, created space and use of the timber structures is hindered by several factors. Firstly, movement into, through and out of the structure was probably dictated by social convention and rules of behaviour, unknown to us (Parker Pearson and Richard, 1994, Blomberg, 1992). Secondly, the organic nature of the timber structures means that we are left with posthole, bedding trenches and if we are lucky, sparse living surfaces. Any less substantial fixtures employed to divide space or influence movement might not have been preserved.

That said, we are also left with material culture. Thomas (2007: 424) has argued that we should be looking at "the way that people inhabit a landscape, and the extent to which new material and symbolic media transformed their existence". Since the timber structures date to the very early Neolithic in Britain, they were probably the setting for early use of the novel new materials like ceramics, domesticates and polished stone tools. The large quantities of grains found at some of the timber structures and the sickles found at Yarnton and Parc Bryn Cegin confirm cereal harvesting was taking place at the time the structures were being used.

What is interesting is the limited quantity and restricted amount of other material reported at every excavated early Neolithic timber structure (Debert, 2010, Sheridan, 2010, Hayden, 2008, Hey and Barclay, 2007, Brophy, 2007, Kenney and Davidson, 2006). A lack of material has been noted at some Neolithic 'ritual' sites, perhaps indicating the area was kept clean of debris or regularly cleaned (Cummings and Whittle, 2004, Topping, 1996, Garrow, 2007, Bradley, 2005). The pristine condition of the lithics that have been recovered from the timber structures suggest that they were not exposed to wear on a surface before deposition. This in conjunction with the fact that lithics and other artefact classes are only recovered from post fills, bedding trenches and the occasional depression may indicate that they were deposited as a result of cleaning. This cleaning would have had to be frequent because, as stated above, the lithics were unworn. Regular cleaning may well indicate that the early Neolithic timber structures served a special function.

I am not suggesting that the timber structures had a strictly ritual function however, I do not have the space to discuss this fully. I will only say that if one thinks about the Maori meeting houses or the Bwayma yam store houses with their interwoven domestic, ritual and social roles, parallels can be drawn with the early Neolithic timber structures (Aoyama, 1994, Aoyama, 2007, Barber, 2003, van Meijl, 1993). As Kirk (2006: 334) argues, "people are able to adopt an almost infinite number of attitudes towards an object, place, building or monument." It would be folly to suggest that the timber structures were viewed in the same manner by: the people who built them, those that used them generations later and those that finally chose to burn them down.

The possible attitudes that could be invoked by these large structures are endless, but the important point is simply that they are not mutually exclusive. The time necessary for their construction, when compared with the time required for the erection of more ephemeral structures, is considerable. It follows that their meaning must also have been greater. Additionally, the structure at White Horse Stone was orientated NNW-SSE, which is perpendicular to the slope and meant that the northern and southern ends were at different heights (Hayden and Stafford, 2006). Therefore, the location, orientation and probably sight lines from the structure were important enough to warrant the extra expenditure of energy. The fact that Mesolithic activity on the sites of the structures is relatively limited, if present at all, in contrast to Neolithic tombs suggests that the meaning of the site did not lie in the history of the place, but possibly in their relation to other sites and/or in the materials gathered and amalgamated within the structures themselves. The mere act of erecting the timber structures changed the locality and the people involved (Tilley, 2007, Winterbottom and Long, 2006, Edmonds, 1999).

Summation

The narrative that could be woven about the timber structures is far greater then space will allow in this paper. As this was only the second of two objectives our account of the timber structures is even more restricted. What should be said is that monuments and now houses, are being seen as part of the process of creating identity and place (Whittle, 1996, Barrett, 1994). As the early Neolithic timber structures are the only architecture being constructed during the initial Neolithic it is here that they can truly inform about the nature of this pivotal period.

In the above narrative the result of the lithic microwear study did not dominate discussions nor should they have been placed ahead of other material evidence. Instead it is hoped this paper demonstrated that microwear can be used to enlighten our interpretations without turning changing the focus to scientific methodology or results. Finally the author suggests that the future of lithic and microwear studies does not lie with the next technical innovation but with theoretical integration and interpretation, now starting to appear in the literature (Debert, 2010, Warren and Dolan, 2007, Rivals and Deniaux, 2005).

References

Andrefsky, W. J. 1998. Lithics: Macroscopic Approaches to Analysis, Cambridge, Cambridge University Press.

Aoyama, K. 1994. Socioeconomic Implications Of Chipped Stone From The La-Entrada Region, Western Honduras. Journal of Field Archaeology, 21, 133-145.

Aoyama, K. 2007. Elite artists and craft producers in Classic Maya society: Lithic evidence from Aguateca, Guatemala. Latin American Antiquity, 18, 3-26.

Barber, I. 2003. Sea, land and fish: spatial relationships and the archaeology of South Island Maori fishing. World Archaeology, 35, 434-448.

Barclay, G., Brophy, K. & Mcgregor, G. 2002a. Claish, Stirling: An early Neolithic structure in its context. Proceedings of the Society of Antiquaries of Scotland, 132, 65-137.

Barclay, G., Brophy, K. & Mcgregor, G. 2002b. Claish, Stirling: An early Neolithic structure in its context. Proceedings of the Society of Antiquaries of Scotland 132, 65-137.

Barrett, J. 1994. Fragments from antiquity: An archaeology of social life in Britain, 2900-1200 BC, Oxford, Blackwell.

Blomberg, B. 1992. Domestic Architecture and the Use of Space - an Interdisciplinary Cross-Cultural-Study - Kent,S. American Antiquity, 57, 738-739.

Bradley, R. 2003. A Life Less Ordinary: the Ritualization of the Domestic Sphere in Later Prehistoric Europe. Cambridge Archaeological Journal, 13, 5-23.

Bradley, R. 2005. Ritual and Domestic Life in Prehistoric Europe, London, Routledge.

Bradley, R. & Edmonds, M. 1993. Interpreting the Axe Trade: Production and Exchange in Neolithic Britain, Cambridge, Cambridge University Press.

Brophy, K. 2007. From Big Houses to Cult Houses: Early Neolithic Timber Halls in Scotland. Proceedings of the Prehistoric Society, 73, 75-96.

Cook, J., Dumont, J. 1987. The Development and Application of Microwear Analysis Since 1964. In: Newcomer, G. D. G. S. M. H. (ed.) The Human Uses of Flint and Chert, 53-62.

Cummings, V. & Whittle, A. 2004. Places of Special Virtue: Megaliths in the Neolithic Landscapes of Wales, Oxford, Oxbow.

Darvill, T. 1996. Neolithic Buildings in England, Wales and the Isle of Man. In: Darvill, T. & Thomas, J. (eds.) Neolithic Houses in Northwest Europe and Beyond. Oxford, Oxbow.

Darvill, T. 2007. Building memories: The neolithic Cotswold long barrow at ascott-under-wychwood Oxfordshire. Antiquity, 81, 811-812.

Debert, J. 2010. Functional Microwear Analysis of Lithics from British Early Neolithic Timber Structures. PhD, The University of Manchester.

Debert, J. & Sherriff, B. L. 2007. Raspadita: A New Lithic Tool from the Isthmus of Rivas, Nicaragua. Journal of Archaeological Science, 34, 1889-1901.

Donahue, R. E. & Burroni, D. B. 2004. Lithic microwear analysis and the formation of archaeological assemblages. In: Walker, E. A., Wenbansmith, F. & Healy, F. (eds.) Lithics in Action, 140-148.

Edmonds, M. 1999. Inhabiting Neolithic landscapes. Journal of Quaternary Science, 14, 485-492.

Evans, A. A. & Donahue, R. E. 2005. The Elemental Chemistry of Lithic Microwear: an Experiment. Journal of Archaeological Science, 32, 1733-1740.

Fairweather, A. D. & Ralston, I. B. M. 1993. The Neolithic Timber Hall at Balbridie, Grampian Region, Scotland - the building, the date, the plant macrofossils. Antiquity, 67, 313-323.

Fletcher, R. 2008. Some spatial analyses of Chalcolithic settlement in Southern Israel. Journal of Archaeological Science, 35, 2048-2058.

Fraser, S. 2006. Digging Deep in Time- Crathes Excavations Go Back 10,000 Years. Archaeology Bulletin, Number 25, 1.

Garrow, D. 2007. Placing Pits: Landscape Occupation and Depositional Practice During the Neolithic in East Anglia. Proceedings of the Prehistoric Society, 73, 1-24.

Garton, D. 1991. Neolithic Settlement in the Peak District: Perspective and Prospects. In: Hodges, R. & Smith, K. (eds.) Recent Developments in the Archaeology of the Peak District. Sheffield: J.R. Collis Publication.

Hardin, J. W. 2004. Understanding domestic space: An example from Iron Age Tel Halif. Near Eastern Archaeology, 67, 71-83.

Hayden, C. 2008. White Horse Stone and the Earliest Neolithic in the South East. South East Research Framework Resource Assessment Seminar, 1-16.

Hayden, C. & Stafford, E. 2006. The Prehistoric Landscape at White Horse Stone, Boxley, Kent. CTRL Integrated Site Report Series Unpublished. London: Oxford Archaeological Unit, Wessex Archaeology, Joint venture for CTRL.

Hey, G. & Barclay, A. 2007. The Thames Valley in the Late Fifth and Early Fourth Millennium cal BC: and the Evidence for Change. Proceedings of the British Academy, 144, 399-422.

Hey, G. & Bell, C. 2000. Yarnton Floodplain B Post-Excavation Analysis Research Design: Modules 3,4,5 and Overview. Oxford Archaeological Unit.

Hogberg, A., Puseman, K. & Yost, C. 2009. Integration of use-wear with protein residue analysis - a study of tool use and function in the south Scandinavian Early Neolithic. Journal of Archaeological Science, 36, 1725-1737.

Kenney, J. 2006. Archaeology at Parc Bryn Cegin, llandygai [Online]. http://www.heneb.co.uk/llandegai. [Accessed 06/01/2010].

Kenney, J. & Davidson, A. 2006. Parc Bryn Cegin Llandygai: Assessment of Potential for Analysis Report.

Kirk, T. 2006. Materiality, personhood and monumentality in early Neolithic Britain. Cambridge Archaeological Journal, 16, 333-347.

Lombard, M. 2005. A method for identifying Stone Age hunting tools. South African Archaeological Bulletin, 60, 115-120.

Lynch, F. & Musson, C. 2004. A Prehistoric and Early Medieval Complex at Llandegai, Near Bangor, North Wales. Archaeologia Cambrensis, 150, 17-142.

Nanoglou, S. 2008. Building biographies and households - Aspects of community life in Neolithic northern Greece. Journal of Social Archaeology, 8, 139-160.

Newcomer, M. H., Grace, R. & Unger-Hamilton, R. 1987. Microwear Analysis, Blind Tests and Texture Analysis. In: Sieveking, G. & Newcomer, M. H. (eds.) The Human Uses of Flint and Chert: Papers from the Fourth International Conference on Flint, 253-263. Cambridge: Cambridge University Press.

Odell, G. 1988. Technical Aspects of Microwear Studies on Stone Tools - Owen,Lr, Unrath,G. American Antiquity, 53, 203-204.

Odell, G. H. & Odell-Vereecken, F. 1980. Verifying the Reliability of Lithic Use-wear Assessments by 'Blind Tests': The Low Power Approach. Journal of Field Archaeology, 7, 87-120.

Parker Pearson, M. & Richard, C. 1994. Ordering the World; Perceptions of Architecture, Space and Time. In: Parker Pearson, M. & Richard, C. (eds.) Architecture and Order: Approaches to Social Space, 1-37. London, Routledge.

Ralston, I. M. B. 1982. A Timber Hall at Balbridie Farm. Aberdeen University Review, 168, 238-249.

Richards, C. 1993. Monumental Choreography: Architecture and Spatial Representation in Late Neolithic Orkney. In: Tilley, C. (ed.) Interpretive Archaeology, 143-178. Oxford, Berg.

Richards, C. 1996. Monuments as landscape: Creating the centre of the world in late Neolithic Orkney. World Archaeology, 28, 190-208.

Rivals, F. & Deniaux, B. 2005. Investigation of human hunting seasonality through dental microwear analysis of two Caprinae in late Pleistocene localities in Southern France. Journal of Archaeological Science, 32, 1603-1612.

Rots, V. & Williamson, B. S. 2004. Microwear and Residue Analysis in Perspective: the Contribution of Ethnoarchaeological Evidence. Journal of Archaeological Science, 31, 1287-1299.

Ruggles, C. & Barclay, G. 2000. Cosmology, calendars and society in Neolithic Orkney: a rejoinder to Euan MacKie. Antiquity, 74, 62-74.

Semenov, S. 1964. Prehistoric Technology, Bath, Adams and Dart.

Shea, J. J. & Klenck, J. D. 1993. An Experimental Investigation of the effects of trampling on the Results of Lithic Microwear Analysis. Journal of Archaeological Science, 20, 175-194.

Sheridan, A. 2007. Neolithic Scotland: Timber, stone, earth and fire. Cambridge Archaeological Journal, 17, 240-243.

Sheridan, A. 2010. The Neolithisation of Britain and Ireland: the Big Picture. In B. Finlayson and G. Warren (eds), landscapes in Transition, 89-105. Oxford, Oxbow.

Thomas, J. 1999. Understanding the Neolithic, London, Routledge.

Thomas, J. 2007. Mesolithic-Neolithic Transitions in Britain: From Essence to Inhabitation. Proceedings of the British Academy, 144, 423-439.

Tilley, C. 2007. The Neolithic Sensory Revolution: Monumentality and the Experience of Landscape. Proceedings of the British Academy, 144, 329-345.

Tipping, R., Bunting, M. J., Davies, A. L., Murray, H., Fraser, S. & Mcculloch, R. 2009. Modelling land use around an early Neolithic timber 'hall' in north east Scotland from high spatial resolution pollen analyses. Journal of Archaeological Science, 36, 140-149.

Topping, P. 1996. Structure and Ritual in the Neolithic House: Some Examples from Britain and Ireland. In: Darvill, T. C. & Thomas, J. (eds.) Neolithic Houses in Northwest Europe and Beyond. Oxford: Oxbow.

Ungerhamilton, R. 1989. Experimental Microwear Analysis - Some Current Controversies. Anthropologie, 93, 659-672.

Unrath, G., Owen, L. R., Van Gijn, A., Moss, E. H., Plisson, H. & Vaughan, P. 1986. An Evaluation of Use-wear Studies: a Multi-analyst Approach. Early Man News, 9/10/11, 117-175.

Van Meijl, T. 1993. Maori Meeting-Houses in and over Time. In: Fox, J. (ed.) Inside Austronesian Houses, 194-218. Canberra, Australian National University.

Vaughan, P. 1985. Use Wear Analysis of Flaked Stone Tools, Tucson, University of Arizona Press.

Wadley, L. & Lombard, M. 2007. Small Things in Perspective: the Contribution of our Blind tests to micro-residue Studies on Archaeological Stone Tools. Journal of Archaeological Science, 34, 1001-1010.

Wadley, L., Lombard, M. & Williamson, B. 2004. The first residue analysis blind tests: results and lessons learnt. Journal of Archaeological Science, 31, 1491-1501.

Warren, G. & Dolan, B. 2007. Warrenfield, Crathes: Stone Tools, Draft Report 1-17.

Wentink, K. 2006. Ceni n'est pas une Hache: Neolithic Depositions in the Northern Netherlands. Amsterdam, Sidestone press.

West, J. A. & Louys, J. 2007. Differentiating Bamboo from Stone Tool Cut Marks in the Zooarchaeological Record, with a Discussion on the Use of Bamboo Knives. Journal of Archaeological Science, 34, 512-518.

Whittle, A. 1996. House in Context: Buildings as Process. In: Darvill, T. & Thomas, J. (eds.) Neolithic Houses in Northwest Europe and Beyond 13-26. Oxford: Oxbow Books.

Winterbottom, S. J. & Long, D. 2006. From abstract digital models to rich virtual environments: landscape contexts in Kilmartin Glen, Scotland. Journal of Archaeological Science, 33, 1356-1367.